FIT2FIGHT

2025 Second and Updated version

A PRACTICAL GUIDE FOR MANAGING THE ASIAN HORNET

Northern Bee Books

A Practical Guide For Managing The Asian Hornet
Copyright © Alan Baxter

First published 2024 by
Northern Bee Books,
Scout Bottom Farm,
Mytholmroyd,
West Yorkshire
HX7 5JS (UK)
Tel: 01422 882751
Fax: 01422 886157
www.northernbeebooks.co.uk

2nd Edition Published by Northern Bee Books 2025

ISBN 978-1-914934-99-5

Design and artwork DM Design and Print

Vespa velutina nigrathorax

For Penny

Contents

Introduction   **1**

3 Pillars of Fit2Fight   **3**

Asian Hornet Description   **4**

Nests   **6**

     Embryo   6
     Primary   7
     Secondary   7

Diet   **11**

Spread   **17**

Threats   **22**

     To honey bees   22
     To diversity   24
     To the economy   25
     To people   26

Preparing the Troops for battle   **27**

     Lifecycle   27
     A Phased response   29

F2F Healthy bees   **30**

     Swarms   31
     Apiary Hygiene   31
     A Clean Environment   33
     Comb Change   34
     Spring cleaning   35
     Routine Inspections   35
     Health Inspections   35
     Notifiable Diseases & Pests   38

AFB 39

EFB 41

Other Common Brood Diseases 43

CBPV 45

DWV 47

Nosema 49

Varroa **51**

Parasitic Mite syndrome 52

Population Growth 54

Monitoring 55

Insurance **62**

F2F Strong colonies **63**

Spring Colony Building 64

Insulation 67

Early Spring Feeding 67

Swarm Prevention & Control 69

Weak Colonies 73

Queen Rearing 74

Strong Colonies Going into Winter 75

Pests & Predators 76

Insulation 77

Summary 77

F2F Well-Fed Stock **79**

Sugar Syrup & Fondant 79

Pollen 84

Water 87

Hot Weather 89

F2F Defence **90**

Apiary Biosecurity 90

Physical Defences 91

Traffic Light System 91

Defence Equipment 93

Entrance Reducers 93

Muzzle & Electric Harp 94

F2F Counterattack   **97**

    Trapping   98
    Types of Trapping   101
    Trap Design   107
    Bait   109
    Recording & Mapping   109

Finding and Destroying nests   **111**

Conclusion   **121**

Annexes   **122**

    Annex A Instructions for Shook Swarm   122
    Annex B Instructions for Bailey Comb Change   124
    Annex C Testing for Nosema   126
    Annex D Instructions for Bailey Comb Change for a Weak Colony 128
    Annex E Instructions for Nucleus Method of Swarm Control 131
    Annex F Instructions for Pagden Artificial Swarm   133
    Annex G instructions for Demaree Method of Swarm Control 135
    Annex H Winter Preparation   137
    Annex I F2F Calendar   141

Abbreviations   **142**

Source material/references   **143**

Acknowledgments   **144**

Introduction

Vespa velutina nigrathorax, the Asian Hornet, is here to stay, but it's not the end of the world.

It's not the end of beekeeping.

But it could be the end of beekeeping as we presently know it.

For our bees to survive the ravages of predation by Asian Hornets, and to continue to thrive, we may need to change the way we keep our bees and alter our expectations of what they can do.

This means taking command in our apiaries, reducing the level and stress of predation, and ensuring that the colony continues to function as normally as possible despite Asian Hornets.

My first experience of Asian Hornets wreaking havoc was in my modest apiary in the Loire Valley of France. I learned a great deal from my friend André, a shrewd, weather-beaten *Maître apiculteur* of a certain vintage.

He taught me that, by putting in place a system of Integrated Apiary Management, we can survive the Asian Hornet problem, get our bees through the predation period and into winter, and still have a crop of honey in return for our efforts.

To achieve this, we have to be very close to our colonies and sharpen up our beekeeping skills and apiary procedures. To put it simply, we need to raise our game and start thinking out of the box.

This book, which is aimed at beginners and more experienced beekeepers alike, proposes a set of simple principles with guidelines for we amateur beekeepers to help manage our bees and apiaries when they are under attack. We aim to ensure that our investment of time, effort and money is still worthwhile in the new situation.
Whilst it is not a textbook on the Asian Hornet, it contains the basic information we need as the starting point for planning our response to this new threat to the health and wellbeing of

1

our colonies of honeybees. The book is focused more on how the beekeeper can respond to the threats it poses in their apiary.

For more detailed information on the Asian Hornet try the Second Edition of Dr Sarah Bunker's excellent *Yellow-Legged Asian Hornet – A Handbook (Psion Press 2024).*

The increasing use of the term *Yellow-Legged Asian Hornet,* a more descriptive term aimed to help with identification, is a bit of a mouthful so the original British name of *Asian Hornet* will continue to be used in this book.

Three Pillars of Fit2Fight

When the future existence of any organism - animal, plant or insect - is under threat it has a better chance of survival if it is fit, numerically strong and well-nourished. With better understanding of the timing of the predation, we can take steps to ensure that the three pillars of Fit2Fight are in place when they will be most effective for our frontline bees. The 3 pillars are:

- Healthy bees
- Strong colonies
- Well-fed stock

We will examine each of the pillars in detail in later chapters, but first let's look at some basic facts about the Asian Hornet, its life cycle, what it eats, the way it spreads and the threats it poses.

It will become clear that, whilst the Asian Hornet is a problem for beekeepers, it's not a beekeepers' problem.

Description

The Asian Hornet reaches an average of 25 mm in length for workers and drones, and 30 mm for queens. These measurements can vary according to the season in which they emerge. Adults emerging in early spring, when forage is less plentiful, tend to be smaller.

Its main distinguishing features are:

▸ Black head with an orange face

▸ Thorax black & velvety

▸ Dark abdomen with wide orange stripe on 4th segment

▸ Yellow tips to legs

HOW TO IDENTIFY THE YELLOW-LEGGED ASIAN HORNET

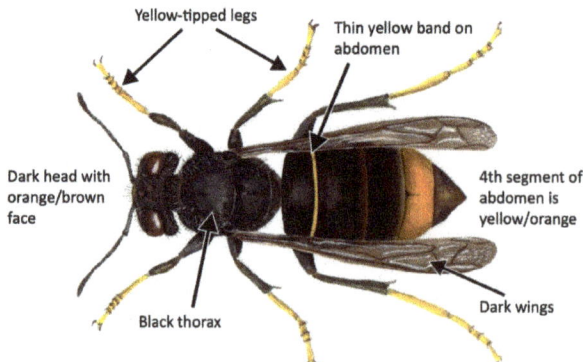

ILLUSTRATION BY SARAH BUNKER, 'THE YELLOW-LEGGED ASIAN HORNET - A HANDBOOK'

Asian hornet features CREDIT Sarah Bunker - The Yellow-legged Asian Hornet A Handbook

To complicate matters, there are a number of different look-alikes:

European Hornet (*Vespa crabro*)

Wood wasp

Hornet mimic hoverfly

Comparison between European Hornet, Asian Hornet and Common Wasp

Nests

The young queen newly emerging from hibernation starts the process of building her colony by setting up a little starter home. As the workers emerge, they progressively take over the role of home builder and forager for the rapidly growing family.

Embryo nest

In spring, after winter hibernation, the queen builds a small **Embryo** nest where she lays her first eggs. Starting with a central stalk or petiole attached to a convenient beam or branch, the queen adds cells around it in which she lays her eggs, with a round protective cover, in a structure about the size of a golf ball. The structure is like papier-mache, made of plant material mixed with saliva and water. The nest is usually hidden in holes or sheltered positions in or near buildings such as sheds, garages, barns or eaves where access to sources of food and water is easy. Holes in trees or walls are ideal and protected places such as bird boxes or outdoor barbecues.

The queen needs to leave the nest to forage for food for herself and for the larvae and must return quickly to keep her growing brood warm. She will continue to forage for 2 to 3 weeks after the first workers emerge, but they gradually take over foraging duties and the queen then remains in the nest.

Her existence at this stage is still precarious and the attrition rate among early emerging queens is high.

Embryo nest showing first cells with eggs.

Primary Nest

As the brood increases the nest grows to accommodate more brood. This larger structure is known as the **Primary** nest and may be a simple enlargement of the embryo nest or a new one on a different site. The primary nest is still usually located in or near buildings and sources of water. As more workers emerge, they gradually take over foraging and nest building duties. The material for the nest is found by scraping different types of hard and soft wood and mixing it with saliva and water to produce layers of different colours and textures giving a striated outward appearance. At this stage the entrance to the nest is still at the bottom.

Primary nest with the entrance at the bottom

Secondary Nest

As more and more workers emerge the colony outgrows the primary nest and moves to a bigger structure known as the secondary nest. This may be on the same site but is more often located in tall trees, in brambles, or hidden underground in a culvert, under a manhole or a cavity in a building. Nests have been found inside gas or electricity meters and in the rooms of unoccupied houses.

The structure of the nest is an impressive feat of architecture and engineering in tiers with galleries of cells in concentric circles, supported by very strong pillars. The roof has additional 'struts' to support the weight of the increasing layers of comb beneath it. Multiple air pockets give the nest good insulation properties.

The entrance to the secondary nest is at the side, often facing east, and some nests have been found to have secondary entrances in the lower part. Other holes in the outer shell resembling entrances are unfinished air pockets and have no function. Often built and anchored around branches at the top of tall trees the nests are able to resist high winds, and although not waterproof they dry very quickly after heavy rain.

The size of the secondary nest varies. In France and Spain for example, completed nests are known to reach up to 1 metre or more in length and 80 cm in diameter, whereas those in the Channel Islands and the UK have so far been much smaller, in the region of 48 cm high by 42 cm wide on Jersey. Despite their size they are well camouflaged and very difficult to see whilst leaves are still on the trees or hidden deep in banks of brambles.

The nest is abandoned at the end of the season when the newly mated queens go into hibernation and the rest of the colony dies.

Secondary nest in a tree. Note the position of the entrance hole on the side of the nest

Roof section of a nest dissected by Bob Hogge on Jersey

Photo by Alan Baxter

Interior of nest showing tiers or galleries
Photo by Alan Baxter

Diet

The Asian Hornet is a voracious apex predator, which means it has no serious competition for territory and food. They are opportunist feeders, taking food wherever they find it, including carrion such as roadkill, open food waste behind restaurants, or stealing from fish, seafood and meat counters in outdoor markets. In France some outdoor markets have moved indoors during summer and autumn due to the high number of insects pestering the stall holders and their customers.

Unfortunately, they are mainly consumers of insects. Their diet depends on where the colony is located, the season and the availability of prey. The following gives an idea of the range of insects consumed. In every case, bees and other pollinators of all types form a major part of the Asian Hornet's diet.

The honey bee in the apiary is particularly vulnerable as there are large numbers of them concentrated in a small space at the time when the hornets' main hunting season is in full swing. Returning foragers, slow and heavily laden with nectar, are easy targets and hornets can be seen hawking round hives whenever nectar and pollen are being collected. Honey bees foraging together in large numbers are ready targets and hornets can be seen hunting wherever there is nectar and pollen being collected.

The unfortunate victims are taken to a nearby tree where they are dismembered to extract the protein-rich flight muscles in the bee's thorax.

Research by a team at Exeter University published in March 2025 highlighted a highly diverse range of insects predated on by Asian Hornets, and that the honey bee is the most frequently predated species.

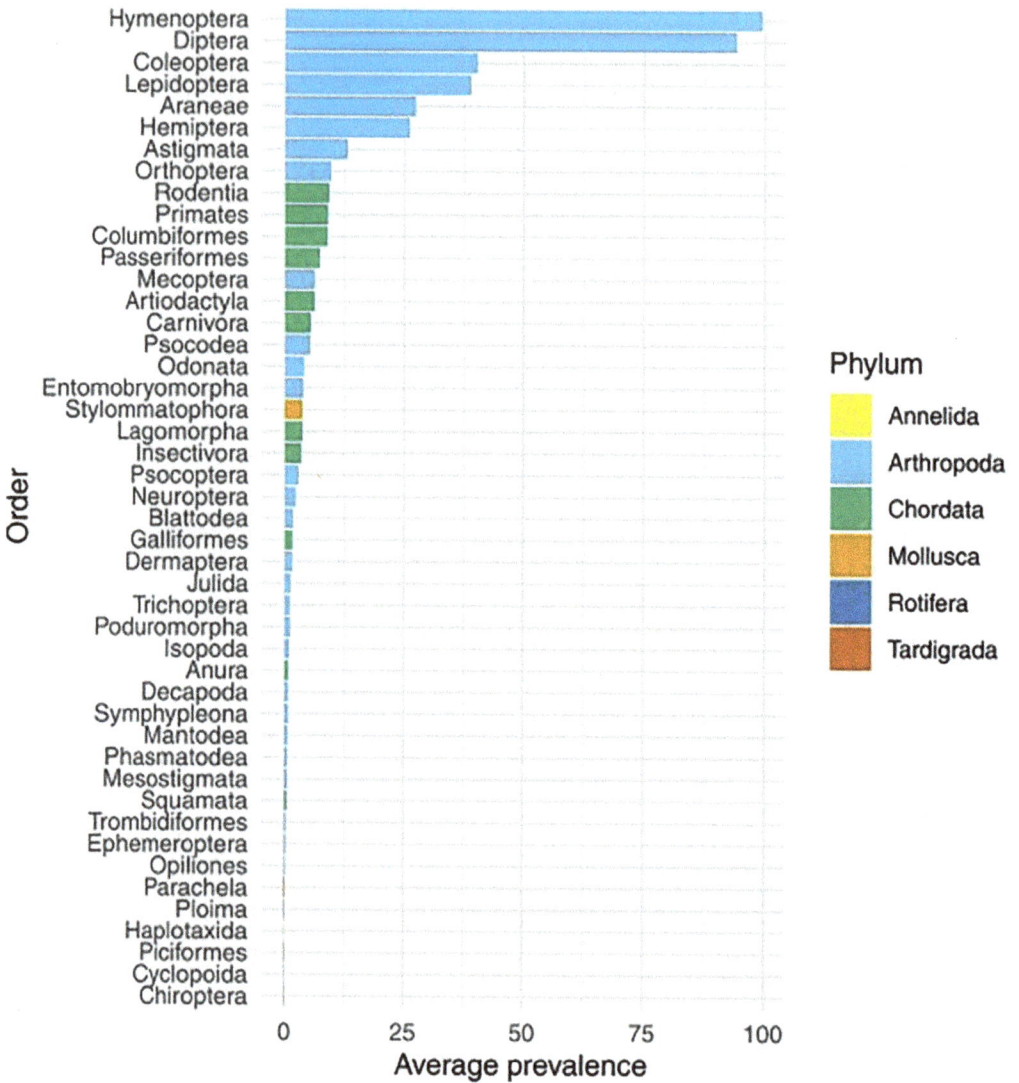

Table 1 showing Hymenoptera as the component of the Asian Hornet diet

Credit Pederson, Kennedy et al

https://www.sciencedirect.com/science/article/pii/S0048969725006138

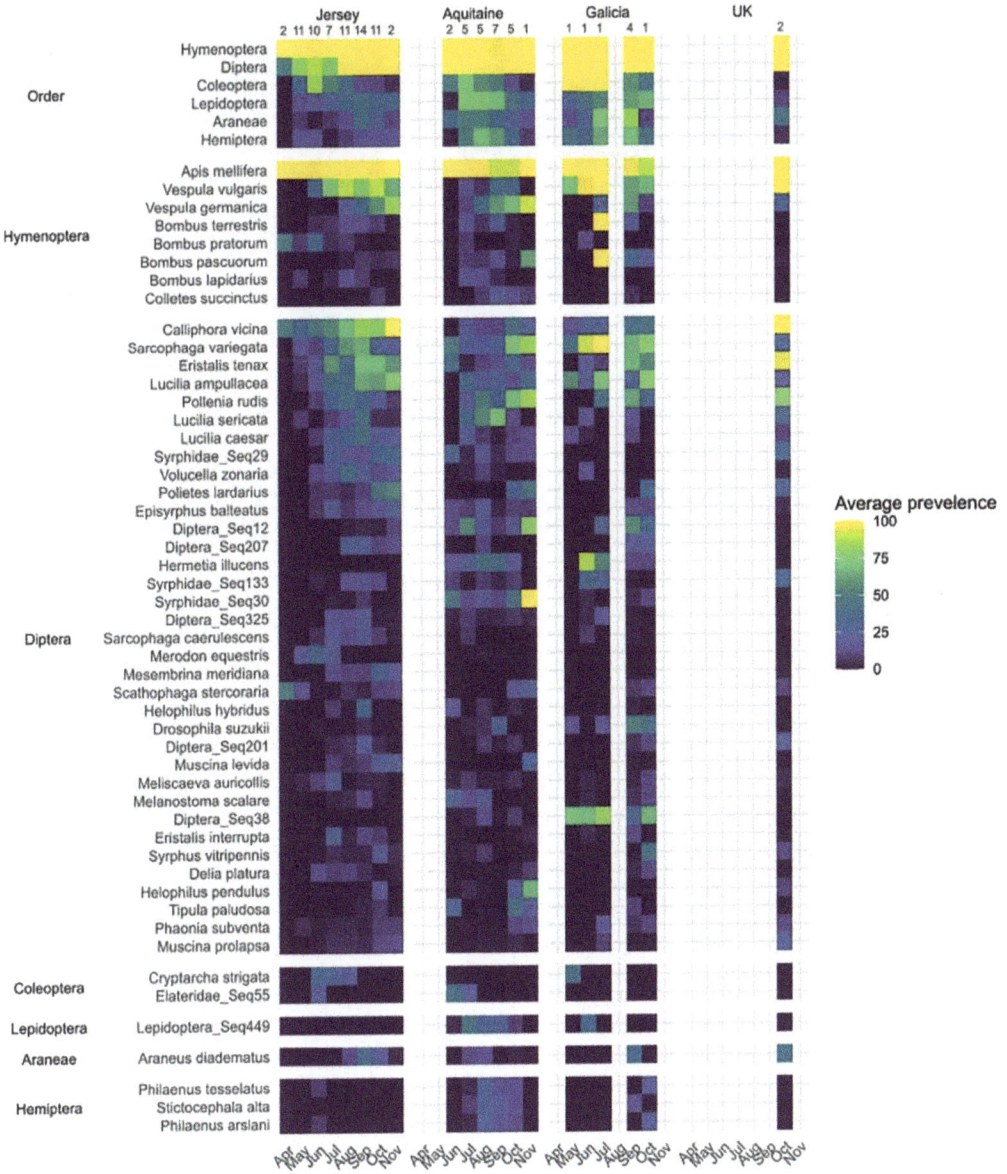

Table 2 showing the honey bee (*Apis mellifera*) in yellow

Credit Pederson, Kennedy et al

https://www.sciencedirect.com/science/article/pii/S0048969725006138

Asian Hornet feeding on camelia in spring

It should be noted that the emerging AH queens in spring, and those preparing for hibernation in the autumn, will forage for sap and nectar as well as protein. They need a supply of the energy to tend their brood and to keep the nest warm in spring. In autumn the queens process nectar and protein to build up the fat bodies which help them survive the winter and sustain them when they first emerge from hibernation the following spring.

Asian Hornet feeding on grapes

Sharing feeding on carrion. The hornet will also take the flies if they're not quick enough!

Foraging workers collect sugar-rich liquids which are often contained in plant sap, and mix them with the protein from their prey into small pellets which they feed to the waiting larvae. In return they receive saliva from the larvae which contains proline for endurance and the amino acids that are processed into sources of energy. There is an exchange of food between larvae and workers ensuring the occupants of the nest of all ages receive the range of nutrients they need, a process in which larvae have a key role producing and storing foodstuffs.

For further reading on diet and nutrition, look no further than *The Yellow-Legged Asian Hornet – A Handbook 2024* by Sarah Bunker.

Spread

Introduced accidentally into France in 2004 in a consignment of garden pots from China, the Asian Hornet has since spread rapidly, becoming established in at least 14 other European countries. The history and scale of the expansion is thoroughly explored in talks given by Andrew Durham which are available online (see reference section at the end of this book).

Once introduced, despite the many dangers it faces, the Asian Hornet population expands exponentially. Once established the invasion progresses through three phases, eradication, containment and long-term management as shown in the graph below:

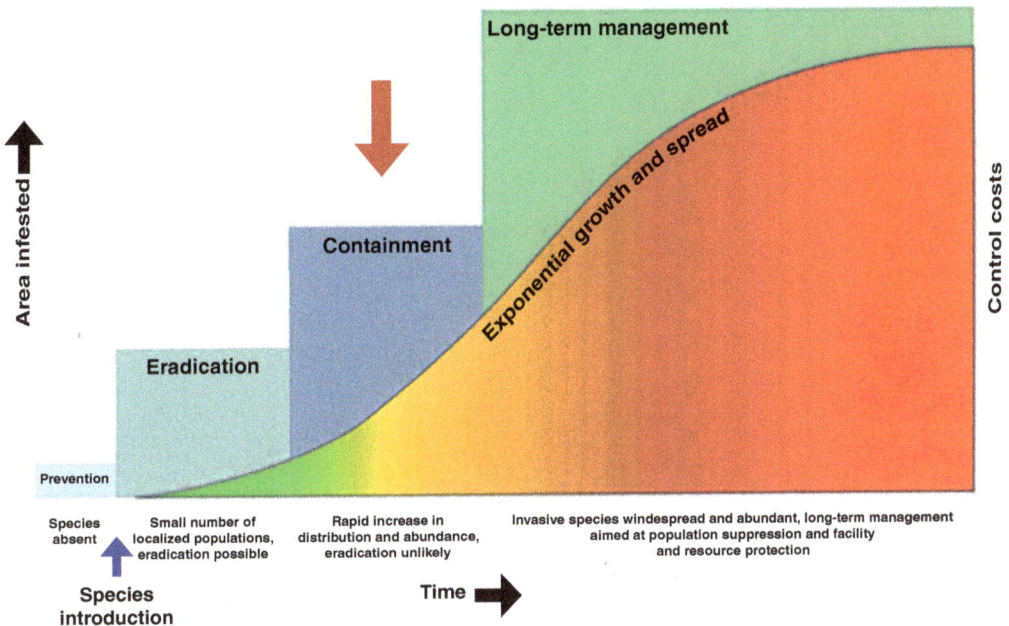

Graph showing the exponential growth of a population.

The time frame can be very short as illustrated in the case of the Wallonia region of Belgium:

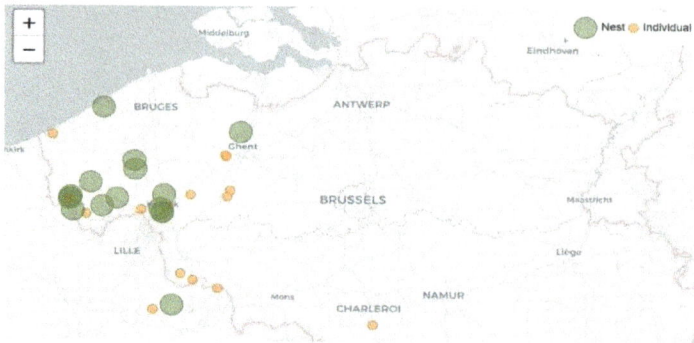

Belgium 2020

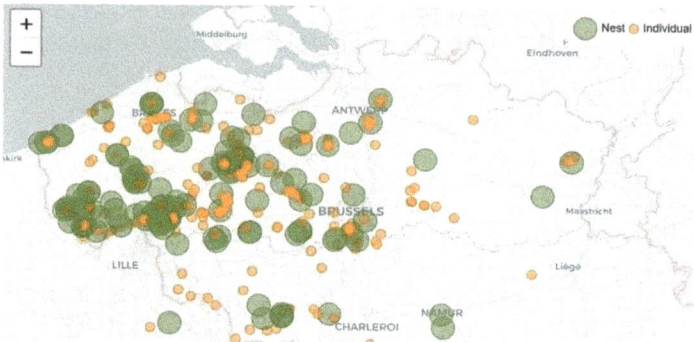

Belgium 2021

Belgium 2022

Climate change is likely to lead to greater expansion of suitable territory in more northerly regions.

Incursions in the UK

Since the first report of an Asian Hornet presence in mainland Britain in 2016 the number of incursions had been very low. However, in 2023 there was a huge increase in numbers, particularly in the south-east of the country. It is speculated that this may have been the result of exceptionally strong winds from the nearby French coast which blew the insects across. The real reason is still not known and is likely to be multi-factorial including favourable weather conditions on both sides of the Channel.

A total of 72 nests were discovered and destroyed leading to optimism that the incursion had been limited to the eradication phase. The discovery in 2024 of a number of nests which are genetically linked to those destroyed the previous year throws doubt on this theory.

Map of England showing the location of Asian Hornet nests (house symbol) or individual insects (eye symbol) in 2023

ASIAN HORNET NESTS 2016 - 2023

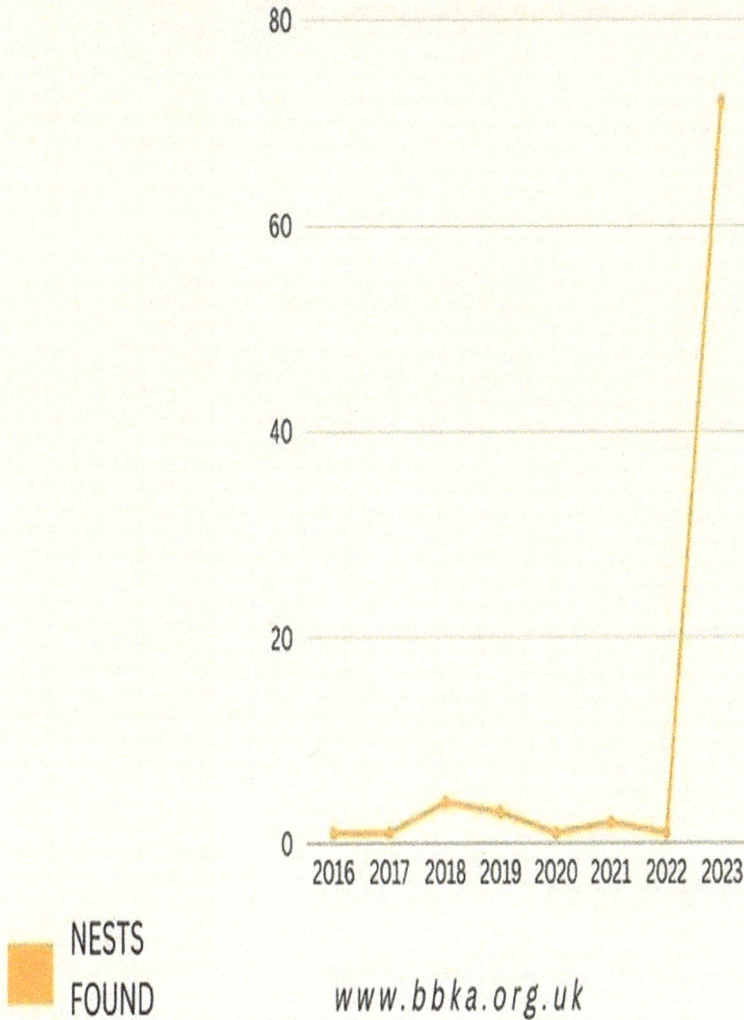

NESTS FOUND

www.bbka.org.uk

Graph showing incursions in the UK 2016-2023

Threats

The arrival of the Asian Hornet poses a number of threats to bees, to biodiversity, to the economy and to public health & safety.

To honey bees

Asian hornets are a threat to all pollinators but as beekeepers we are mainly interested in the potential impact on our honey bees.

Asian Hornets hunt on the wing by hovering ('hawking') in front of the hives, usually with their back to the entrance. They have a wide view of returning foragers, which are easier to catch as they fly more slowly when laden with nectar, pollen, propolis or water. The hornets in flight emit a distinctive loud, deep buzzing sound, the vibrations of which may warn the bees of the impending danger. Pheromones emitted by the hornets soon alert other hornets to the opportunities presented by the large numbers of prey concentrated in the small area of the apiary.

The unlucky victim is carried to a nearby branch and swiftly dismembered. The big protein-rich flight muscles in the bee's thorax are removed and taken as a pellet back to the hornet's nest where it is fed to the hungry larvae.

Asian Hornet dismembering a honey-bee
Image courtesy of Michael Judd

When a colony is under sustained attack the guard bees cluster at the entrance in a defensive wall. If the colony is strong, the hornets hesitate to land near the entrance and avoid entering the bees' territory and any confrontation with the colony's guard bees. So far, so good.

However, the terrified bees stop foraging for nectar, pollen, and water in what is known as foraging paralysis. They become stressed, exhausted, and less able to defend the hive. Their lives are shortened, and they become more susceptible to disease. As normal hive activities cease, the queen is likely to stop laying which, together with the lack of foraging, means the colony will have insufficient winter bees and stores to survive until the following spring.

Colony under attack, bees clustered in defence of the hive entrance.
Image courtesy of Michael Judd

23

The colony becomes progressively weaker, there are no guard bees left to defend the entrance, and the emboldened hornets are free to enter the hive and plunder what is left. Unlike wasps, the Asian Hornet takes everything: bees, honey, and brood. Colony losses in France and Spain as a result of Asian Hornet predation are claimed to be in the region of 30-50%, although some of these can be multi-factorial including loss of forage, extreme weather and pests or diseases.

The main predation period is in mid to late summer and autumn with the maximum pressure on the colony in September. The actual timing depends on geographical location, climate, and the weather at the time. It usually begins in July and has been known to extend into November in some areas. This is the time when we need to be at maximum vigilance. The most intense foraging activity tends to be in the middle part of the day between 1 and 2 pm, although in very hot weather the hornets prefer to hunt earlier or later when it is cooler. They don't fly at night so there is at least some respite for the bees.

We can support our bees by adjusting the timing of the beekeeping calendar, completing as many tasks as possible in the colonies before predation begins and creates stress with all its consequences. The colonies can then be put into 'lockdown' with the entrances reduced to 5.5 mm until the danger has passed.

NB If a queen is damaged or lost during an inspection, once predation has started there is little chance of the colony being able to replace her.

Threat to diversity

An Asian Hornet nest is estimated to consume about 11kg of insects in a season. To give an idea of what this represents:

Average weight of an insect	3mg
1 kg	1,000,000 mg
11kg of insects is	11.000,000/3 = <u>3.67 million insects</u>

In a country like Britain, which already has a severely depleted insect population; (how often is your car windscreen splattered with midges in summer these days?) If it becomes established and widely distributed it will have a huge impact on everything that depends on insects. This includes not only pollination and the control of some pests that are eaten by wasps and ladybirds for example, but also the loss of food for the wild bird population and everything else in the food chain.

Threat to the economy

Among the many economic impacts, the following industries will suffer a major hit when the AH is established:

▸ Agriculture, horticulture, viticulture, beekeeping.

▸ To all major land users including water companies, MOD, large country estates.

▸ Angling, hunting

▸ Leisure and tourism

▸ Costs:

    ▸ Pest control

    ▸ Health

    ▸ Emergency services

    ▸ Assurance and inspection

    ▸ Detection and monitoring

Threat to People

The Asian Hornet is well equipped for survival. Whilst it is relatively passive when foraging away from the nest, it defends its home, its queen and its young ferociously if the nest is disturbed. The first attacking insect emits an alarm pheromone which summons its nest mates to attack the intruders in large numbers with sometimes fatal outcomes.

With a powerful armoury the Asian Hornet is a formidable defence machine:

▸ It is armed with a stinger 3.0 to 3.5mm long.

▸ The stinger is smooth and can penetrate several times.

▶ The sting is reported to be more painful than that of a wasp.

▶ The venom can trigger serious allergic reaction causing acute kidney injury and failure, especially in people who are allergic to stings.

▶ 20 stings or more lead to a high risk of death.

▶ It also emits a liquid from its rear end that, if squirted into the eyes, causes severe discomfort and inflammation.

Asian Hornet stinger
Photo John de Cartaret

An attack is likely if the nest is disturbed within 5 metres and is easily triggered by loud noise and vibration, for example by a hedge-trimmer. Nests hidden in culverts, under manholes or in meter boxes can unknowingly be disturbed. It has been found that in most cases the defenders don't pursue an intruder for more than 10 metres from the nest, although the odd one may persist for longer. It is recommended that the best action to take if you are unlucky enough to be in this predicament is to cover your head and run for your life.

Preparing the Troops for Battle

As in all emergency situations such as natural disasters, pandemics and wars, careful preparation and planning are the essential keys to success and to survival. The Asian Hornet incursion is akin to an enemy invasion, so a military-style approach has been adopted in this book.

We begin by getting to know how the enemy ticks, describe the need for a suitably phased response, then explore in more detail the three pillars of Fit2Fight. We go on to look at our array of defences, then consider the offensive weapons at our disposal. We conclude by describing the actions that can be taken to help limit the spread of the Asian Hornet population.

Life cycle

To understand how we can successfully react in a timely and proportionate way to the threat posed by Asian Hornets to our precious bees, it is essential to consider how they live.

This simple diagram shows the key stages in the life of the Asian Hornet. The traffic-light colours correspond to the level of threat to honeybee colonies and to our responses:

Queens hibernating

Death of colony
New queens into
hibernation

Foundress queens
emerge Feb-May

Build embryo nest.
Eggs laid

Reproduction period

First workers emerge

Predation
period

Honey bees at
risk

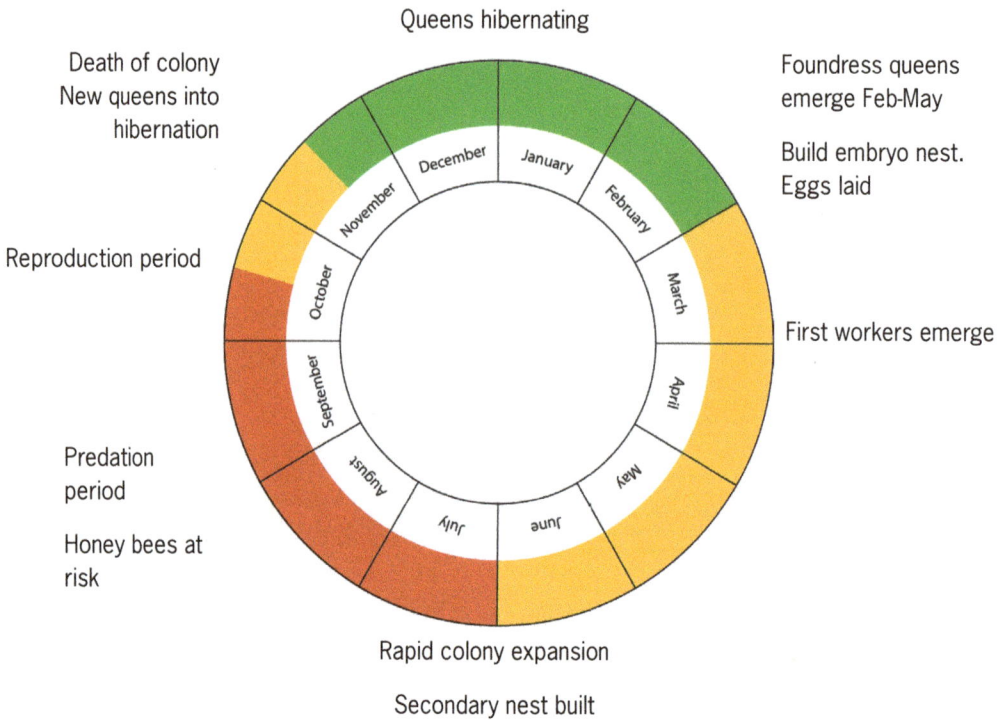

December
January
November
February
October
March
September
April
August
May
July
June

Rapid colony expansion

Secondary nest built

Asian Hornet Life Cycle*Diagram by Alan Baxter*

A Phased Response

"Victory can be foretold when the general knows the time to fight and when not to fight."
Sun Tzu: The Art of War

We can give the bees a better chance of riding the storm by adjusting the timing of our beekeeping calendar to complete as many tasks in the colonies as possible before predation begins, then progressively locking down the hives until the danger has passed. The same traffic light coding as in the Asian Hornet life cycle can be used in our response:

Green   I'm in an area where no hornet nests have been found yet.

Amber   Asian Hornets confirmed in my area.

Red   Asian Hornets have been seen actively hunting in or near my apiary.
In later chapters we will see how this system is put into practice, but first we'll look in detail at the 3 Pillars of Fit2Fight.

F2F Healthy Bees

"A healthy soldier is a ready soldier."
Lt Gen Scott Dingle

As beekeepers we ask thousands of wild creatures to live in artificial wooden or plastic boxes, close to the ground, a far cry from the holes in trees or rock faces high above the earth of their own choosing. This brings a number of heavy responsibilities, not least of which is for the health and wellbeing of the livestock in our care. Honey bees are susceptible to a wide variety of pests and diseases which can kill them or shorten their lives leading to total collapse of the colony and of those colonies with which they come into contact.

By living close to other colonies in apiaries, a situation that doesn't occur in the wild, we create an opportunity for sickness or parasites in one colony to easily be passed to the others. In the course of their wide-ranging foraging activities, honey bees meet or forage on the same sources as other bees where they can easily leave, or pick up, infections or parasites. At certain times of the year, especially after the summer flow, robbing between hives and apiaries is common, and any time of the year foragers can drift into other hives, potentially carrying with them pests and diseases.

We cannot control what happens in the field, but it is vitally important that we do everything we can to prevent the introduction and spread of infections and parasites in our apiaries, especially when they may become the target of predation by Asian Hornets.

In this chapter we will describe the measures we can take to maintain our colonies in good health and minimise the risk of them transmitting problems to other colonies in the apiary and elsewhere in the area. We then describe some of the adult bee and brood diseases that we need to look out for.

In a separate section, we'll consider the impact of Varroa on our colonies and the measures we can take to reduce the Varroa population.

Swarms

We will start with swarms as many of us begin our beekeeping adventure with a swarm of bees from our local beekeeping association. As we progress in our beekeeping journey, we may be lucky enough to collect a swarm ourselves. Usually, we don't know where the swarm came from, what its temperament is like, or what state of health it is in. It may be very aggressive, have a contagious disease or be carrying pests.

If you are given or collect a swarm it should be kept away from any other colonies until after a complete brood cycle in case it is infected with any pests or diseases.

The swarm should not be fed for the first 3 days. This allows them to digest and excrete, outside the hive, any infected food carried from the old brood nest.

During this observation period, treatment with Oxalic Acid can be given before any sealed brood is present, to reduce the varroa level. At the end of quarantine, a careful health inspection should be carried out to confirm the colony is free of disease and parasites. See the chapters below on health inspections, diseases and Varroa.

Apiary Hygiene

Unfortunately, many diseases are carried and transmitted by beekeepers. A vital part of keeping our bees healthy is maintaining the apiary and all our equipment clean and free from contamination. Here are some measures you can take to reduce the risk of introducing or spreading disease:

▶ Do not allow used comb, eggs, larvae or bees from other apiaries to be introduced into your colonies.

▶ Second hand equipment should be scraped and sterilized before use.

▶ If you wear gloves for inspections, reduce the transmission of pathogens between colonies by using disposable ones and changing them between each hive. If you wear Marigolds rinse them in soda crystals and soap solution between hives.

▶ Leather gloves are clumsy, can aggravate the bees, and are a rich breeding ground for infection so are best avoided. If you need to wear leather gloves for protection, you probably need to requeen the colony to have calmer, less defensive bees. If you are allergic to stings and you choose to wear leather, you can put a pair of disposable gloves over the leather ones and change them between hives. Alternatively, more hygienic rubber or latex sting-proof gloves are available.

▶ When brushing off bees, use a soft twig with leaves from a nearby bush. Bee brushes can become clogged with hive debris and carry infection.

▶ After apiary visits wash your bee suit and other clothing in a soda crystals and detergent solution of 1kg soda crystals to 4.5 litres of water.

▶ Hive tools should be scrubbed with wire wool to remove wax and propolis, then rinsed in soda crystals or bleach solution.

▶ The bellows of smokers can be protected from dirt with a disposable shower cap. Smokers can be cleaned by immersion in soda crystal solution and scrubbing with wire wool.

▶ Hive debris such as wax and propolis scrapings should be put in a lidded container and remove from site.

▶ The ground should be kept clear of any debris, especially edible matter such as fallen fruit.

▶ Take care when feeding sugar syrup not to spill any and attract robbers.

▶ Honey supers should be labelled and returned to the same hive for cleaning after extraction.

▶ Used wooden equipment should be scraped free of wax and propolis and sterilized with a blow torch before being stored or reused.

▶ Poly equipment can be scrubbed with wire wool and immersed in a mixture of soda crystals, detergent and hot water.

▶ Super or brood frames should be put in the freezer for 48 hours at -18°C or fumigated with 80% Acetic acid to kill Wax Moth, *Braula coeca* and other parasite eggs, pupae, or pathogens that may be lurking in the comb.

A clean environment

Honey bees furnish their homes with honeycomb made from wax secreted from their own glands, and which they fashion into perfect hexagonal structures. In the comb they raise their young, process and store their food and live for most of their lives. The comb gets a lot of use and the bees, through their hygienic instincts, keep it as clean and in good repair, as they possibly can.

The comb is subject to a lot of wear and tear and doesn't last forever. Dirty brood comb can contain an accumulation of old larval skins and faeces, pathogens, residual chemicals from treatments given by the beekeeper and brought in by foragers, and more. It's the perfect breeding ground for all sorts of nasty pathogens.

In the wild, bees move house regularly by swarming and don't usually stay in one place long enough for the comb to wear out and become unhealthy, whereas we expect them to live in the same hive for years. To give the bees a clean home to live in, and keep them healthy, we change the comb regularly as a routine. This is normally recommended to occur every 3 years, but in the new normal, with the threat of the Asian Hornet hanging over us, we need to increase the frequency of the comb change.

Methods of comb change vary from an incremental system, where a proportion of the comb is changed every year, to a complete change in one go.

Shook swarm performed annually or two-yearly followed by treatment with Oxalic acid, can be a multiple winner by:

▸ Removing accumulated pathogens and chemical residue from treatments and imported when foraging.

▸ Providing the adult bees and larvae with a clean, healthy environment.

▸ Boosting population growth.

▸ Preventing swarming.

▸ Reducing varroa load.

Shook swarm is usually carried out on strong colonies in spring and at a time when there are enough young bees 10-20 days old to draw out the foundation. As it involves loss of brood, it is better to avoid times when the eggs that will develop into foragers for the nectar flows are being laid. Colonies are said to enjoy a period of very rapid growth after this procedure.

For instructions on how to perform a shook swarm please see Annex A.

Shook swarm
Bees are shaken from the original box on the right of the picture onto new foundation on the left.
Courtesy the Animal and Plant Health Agency (APHA), Crown Copyright

Bailey Comb Change is carried out on strong colonies in spring, is a relatively simple procedure, and doesn't involve the loss of brood. Details on this manipulation are at Annex B.

Spring cleaning

At the beginning of the beekeeping season, once the weather is warm enough, we give the hives a spring clean, with a clean floor, brood box and crown board, as well as brushing the debris from under the stand. The supers coming out of store will have been cleaned and treated for wax moth the previous autumn.

Routine Hive Inspections

Every time you open a colony, do so for a specific reason. If you don't have a reason other than impatience or curiosity, leave them alone.
In the beekeeping season routine inspections are carried out weekly. During the inspection establish the following basic information:

▸ Is the queen present, either seen or evident by the presence of eggs and very young larvae?

▸ Has the colony built up since the last inspection, as indicated by the number of frames of brood and seams of bees?

▸ Are there any signs of disease?

▸ Does the colony have enough space for the queen to lay, for incoming stores, and for all the workers at night or when there is no foraging?

▸ Does the colony have sufficient stores until the next inspection?

Health Inspections

More detailed inspections for the presence of diseases or parasites affecting adult bees and brood begin in spring soon after the first inspections of the season and, in the Fit2Fight system, continue monthly until the beginning of predation in mid-late July.

▸ Choose a warm, fine day when many bees are out foraging.

▸ Prepare to take photos and keep a record of each inspection.

▶ Watch the bees at the hive entrance, and inside the hive, and look out for any abnormal appearance or behaviour.

▶ Find the queen and put her in a cage for safe keeping.

▶ Shake the bees off frame by frame and examine the brood carefully, looking out for any abnormality.

Learn to recognize what healthy brood looks like:

▶ Eggs should be laid singly in the gravity bottom of the cells.

▶ The larvae should be creamy-white, c-shaped and with the segmentation clearly visible.

▶ Capped brood should be dry, slightly domed, biscuit in colour, evenly distributed, and with no perforations.

During each health inspection we are looking for signs of unhealthy bees and brood such as:

▶ Dead or dying bees on the ground outside, on the landing board or on the hive floor.

▶ Fighting at or near the entrance.

▶ Yellow or brown stains on the outside of the hive.

▶ Unpleasant smells.

▶ Malformed workers including bees with stunted bodies and deformed wings.

▶ Bees with shiny, black bodies.

▶ Lethargic bees unable to fly or not reacting to smoke.

▶ Scattered brood pattern, sometimes referred to as pepperpot brood.

▶ Multiple eggs on the side of cells.

▶ Knobbly dome-shaped cappings.

▶ Perforated sealed brood with ragged edges.

▶ Sunken cappings.

▶ Greasy looking cappings.

- Scales on the bottom of cells.

- Malformed larvae, not c-shaped and segmented.

- Discoloured larvae (not pale creamy-white).

Examining a frame of brood. Note uneven patchy pattern with missing brood.
Courtesy the Animal and Plant Health Agency (APHA), Crown Copyright

A frame of healthy brood. Even brood with no empty cells and drones at the periphery
Courtesy the Animal and Plant Health Agency (APHA), Crown Copyright

If anything unusual is seen or suspected, begin remedial measures straight away as detailed in the BBKA Healthy Hive Guide* or on Beebase.

Don't hesitate to seek advice from a Bee Inspector if you're not sure what to do.

Notifiable diseases and pests

In Britain there are 2 notifiable brood diseases:

▸ American Foulbrood

▸ European Foulbrood

and 3 notifiable pests that are currently not present in the UK:

- ▸ Tropilaelaps
- ▸ Small Hive Beetle.
- ▸ Varroa destructor is also notifiable but, as it is endemic, it is assumed always to be present.

NB The Asian Hornet is a reportable pest under current regulations but may be elevated to notifiable in the future.

American Foulbrood (AFB) is caused by a spore-forming bacterium called *Paenibacillus larvae* and affects the larva after its cell is capped. A small number of spores are passed to the young larvae in their food and quickly multiply into millions. The larva is overwhelmed and dies of toxaemia or septicaemia in the pupal stage. The dead larva eventually dries into a hard scale stuck to the bottom of its cell which the bees cannot remove.

It is considered to be a very serious disease all over the world and strict statutory control measures are in place in England and Wales.

What to look for:

- ▸ Sunken, wet or greasy-looking brood cappings.
- ▸ Cappings may be pierced by the bees investigating the problem.
- ▸ Scattered brood pattern where the queen refuses to lay.
- ▸ The dead larva pulls out into a rope when explored with a matchstick.
- ▸ Dark, hard scales on the bottom of the cells that the bees cannot remove.

What to do about it

If you suspect you might have American Foulbrood you **must immediately**:

- ▸ Stop the inspection.

▸ Close the hive entrance to one bee space until the bees have stopped flying, then close it completely.

▸ Call the Bee Inspector.

▸ Put the apiary in voluntary stand fast until the Inspector arrives. This means no movement of bees, honey or equipment.

▸ Carefully dispose of disposable gloves, disinfect others with a strong solution of washing soda.

▸ Sterilize your hive tool and other equipment with a strong solution of washing soda and bleach.

▸ Wash your bee suit and contaminated clothes and footwear in washing soda crystals and detergent solution.

The Bee Inspector will confirm the diagnosis with a lateral flow test and possibly send a sample to the laboratory for analysis. They will order the treatment, if any. In the case of AFB, the destruction of the colony is the only option. The apiary remains in stand fast until the Bee Inspector certifies that it is clear of infection.

American Foul Brood (AFB)
Greasy, sunken cappings. Patchy brood. Matchstick used to demonstrate roping of decayed larva.
Courtesy the Animal and Plant Health Agency (APHA), Crown Copyright

European Foulbrood (EFB)

EFB is caused by a bacterium called *Melissococcus plutonius* and affects the larva before its cell is capped. It is spread from the mouthparts of adult bees, given to young larvae in their food and passed between the adults during grooming, food sharing and communication. It can lay dormant and undetected in a colony or comb for many years. In the larva it is activated and reproduces very rapidly, competing with the larva for nutrition. In many cases the larva dies of starvation, but in a strong colony with plenty of food it may survive and continue to develop into an adult. However, the adult bee will be weaker and infected with the disease which it then passes on to the rest of the colony. When death occurs it's usually before the cell would have been capped. The larva dries to a soft, leathery scale which the bees remove, making it less obvious that there is a problem.

What to look for

▶ Discoloured, yellow-brown larvae lying in an unnatural position in the cell.

▶ Melted appearance of the larva with a white line or mass showing through the skin.

▶ The dead larvae do not form a rope when pulled out with a matchstick.

▶ Dead larvae dried to a loose brown scale that is easily removed.

What to do about it

Stop the inspection immediately and carry out all the immediate actions described for AFB above. The Bee Inspector will confirm the diagnosis and decide on the next steps to take.

Mild cases of EFB may be treated by shook swarm, but this is by no means guaranteed to succeed. In severe cases destruction may be the only option. In the past, treatment with the antibiotic Oxytetracycline was given but this is no longer the usual practice in the UK. The Bee Inspector is the sole arbiter in any decision regarding treatment or destruction.

European Foul Brood (EFB)
Discoloured yellow-brown larvae, decaying to leave rubbery brown scales.
Courtesy the Animal and Plant Health Agency (APHA), Crown Copyright

Other commonly occurring brood diseases include Sac brood and Chalk brood:

Sac brood

This disease is caused by a virus of the same name which attacks the pupa and inhibits it from completing its final moult.

What to look for

The leathery skin of the dead pupa is full of fluid, and it lies in its cell in a characteristic gondola or Chinese slipper attitude. It dries into a loose scale which is easily removed by the house bees which then pass on the infection.

What to do about it

In most cases the level of infection is low, and the bees keep it under control. More severe cases can be treated by feeding sugar syrup or requeening.

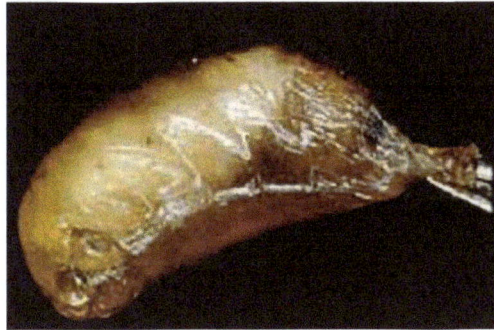

Larva with Sac brood virus
Note the typical 'Chinese slipper' attitude of the swollen, fluid-filled larva.

Chalk brood

Chalk brood is one of the conditions most beekeepers are likely to see at some stage in their beekeeping career. Caused by a fungus called *Acosphaera apis* it usually occurs at times of stress on the colony, for example during prolonged periods of bad weather or in spring when there are insufficient bees to look after the growing number of larvae. The fungus grows best at a temperature of 30 deg C, lower than the optimum brood nest temperature of 34 deg C, and in the presence of higher levels of CO_2.

What to look for

Chalkbrood affects the sealed brood and often appears as perforated cappings. Underneath the cappings the dead larvae are white with yellow or pink mouthparts and a chalky appearance, turning from white to grey and sometimes black. As the dead larvae dry, they form loose 'mummies' which the bees remove and may be found on the hive floor or on the ground near the entrance.

What to do about it

The condition generally disappears once conditions improve, and the bees are able to keep the brood nest warm. However, the spores formed by the fungus can live for up to 15 years in the comb, so a comb change by shook swarm or Bailey comb change is advised in cases of persistent infection. Requeening, hopefully with a more resistant strain, is another option in severe cases. Providing winter insulation and not locating hives in cold, damp places or under a heavy canopy of trees all help to avoid the conditions in which Chalkbrood thrives.

Chalk brood showing typical hard 'mummies' that have been cleared out by the workers.

Other adult and brood diseases which pose a threat to the health and well-being of honeybees are:

Chronic Bee Paralysis Virus (CBPV)

This serious viral infection is present in all countries where Apis mellifera is kept and may also be seen in other Hymenoptera. It can spread like wildfire especially in large, crowded colonies and rapidly lead to colony collapse.

Whilst two forms of the disease have been identified, both are often present at the same time:

Type 1

The most common version of the disease.

What to look for

▶ Stunted bees with bloated abdomens.

▶ Dislocated wings.

▶ Inability to fly.

▶ Signs of dysentery.

▶ Lethargic bees that do not react to smoke.

▶ Dwindling population.

▶ Large number of dead bees on the hive floor and outside the entrance.

Type 2

Sometimes referred to as Black Robber Disease or Hairless Black Bee Disease due to the appearance and behaviour of the infected bees.

What to look for

▸ Bees with shiny, black, hairless bodies

▸ Bees able to fly in the early stages.

▸ Trembling, disoriented bees.

▸ Aggressive behaviour or nibbling of the victims by the non-infected bees.

▸ Guard bees refusing reentry to infected foragers.

▸ Large number of dead bees on the hive floor and outside the entrance.

Crowded conditions in the colony with bees rubbing against each other lead to hairs being pulled out of the bees' abdomens. The resulting wound allows the virus to enter and cause infection. This can often occur at times when the population is growing rapidly.

What to do about it

Overcrowding is a key factor in the spread of the disease and entire colonies can quickly be wiped out.

▸ Give plenty of room by adding supers or an extra brood box.

▸ Raise the brood box by adding a super or an eke underneath so the unaffected bees are not in contact with the dead or dying bees.

▸ Find and cage the queen for safe keeping and carry the brood box away from the stand.

▸ Tip the brood frames onto a large sheet of cardboard.

▸ Scrape and scorch the empty brood box and replace it on the stand.

▸ Shake each frame in the air to remove all the bees and replace the frames in the brood box. The unaffected bees will fly back to the brood box.

- ▶ Repeat this exercise after 6 days to remove the previously infected but symptomless bees.

- ▶ Reassemble the hive and return the queen.

- ▶ Clear up and burn the dead bees and the cardboard.

- ▶ If possible, relocate the hive away from others until the infection has cleared up.

Worker bees demonstrating signs of Type 2 Chronic Bee Paralysis Virus (CBPV)
Shiny, black, hairless bodies with K-Wings
Courtesy the Animal and Plant Health Agency (APHA), Crown Copyright

Deformed Wing Virus

This is a commonly occurring disease, usually associated with a high varroa load. It is persistent at low levels in healthy colonies where the bees' immune systems are able to keep it under control. Once their immune systems are compromised by the presence of large numbers of mites, the pupa become infected and die. Adult bees develop clinical signs of the disease, and the colony can dwindle and die out.

What to look for

Bees crawling about, unable to fly, with stunted bodies and shrivelled wings. You may see varroa mites clinging to their abdomens feeding on the fat body. Their lives are short, and they contribute little to the life of the colony.

What to do about it

There are no known treatments for honey bee viruses and prevention is the best cure. This virus is linked to high varroa loads so treating for mites and keeping levels low is the key to dealing with DWV. Once the varroa population is brought down feeding with thin sugar syrup to give them a boost and introducing a young productive new queen can also help.

Worker bee showing symptoms of Deformed Wing Virus (DWV)
Courtesy the Animal and Plant Health Agency (APHA), Crown Copyright

Nosema

Nosema is a spore-forming, single-cell parasite that affects the digestive system of the adult bee. The fungi multiply in vast numbers in the bee's gut preventing the bees from digesting food, especially pollen. The spores are excreted in the bee's faeces and the combs can become infested with the spores particularly in times when the bees are confined to the hive. The result is a shortened life span leading to dwindling population and eventual death of the colony. Queens are often superseded in late autumn when a new queen has little possibility of being mated.

There are two forms of the organism, *Nosema apis* and *Nosema ceranae*. *N. apis* is more evident in winter and spring, often appearing to clear up during the summer months when infected bees defaecate, and sometimes die, outside the hive. *N. ceranae* can cause severe symptoms at any time of the year and is considered a more virulent form of the disease. It has become widespread in Britain in recent years.

What to look for There are often no obvious signs of the disease, although it may be accompanied by dysentery, in which case yellow or brown stains can be seen on the frames and the outside of the hive. However, dysentery is not always indicative of Nosema as it can be caused by other unrelated problems. Previously strong colonies that fail to survive the winter, or to build up in the spring, are often found to have been infected with Nosema.

Dysentery stains on the outside of the hive and on the frames.

Dysentery is not always present in cases of Nosema and can have a number of other causes.

Diagnosis is by laboratory analysis, but you can carry out a rough test as follows:

▶ Take a few young bees from the centre of the brood nest, put them in a plastic bag or a jar and euthanize them in the freezer or with Isopropyl alcohol.

▶ With forceps or tweezers pull out the intestines from where they exit the body near the sting area.

▶ The midgut, which is normally brownish in colour in a healthy bee, is white and often distended in the infected bee.

To confirm the diagnosis, take a sample of 50 bees from the top of the brood nest, euthanize them in the freezer or with Isopropyl alcohol and send them to a laboratory for analysis.

If you have a compound microscope, please see Annex C for instructions on this procedure.

What to do about it

There are currently no approved treatments for Nosema in the UK. Instead of using medicines for treatment of Nosemosis, beekeepers should try to maintain their colonies in good health by applying sound husbandry practices such as maintaining strong, well fed and disease tolerant colonies, headed by young and prolific queens.

Replacing the infected comb by **Bailey comb change for a weak colony** can bring some relief, and feeding with invert syrup may help. Testing for Nosema in autumn followed by remedial action can, with luck, reduce the risk of winter or spring colony loss.
For instructions on how to carry out this manipulation please see Annex D.

Varroa

Varroa destructor is aptly named. This horrible little beastie (technically an Ectoparasitic mite) is a major cause of failure in our colonies and is associated with a wide variety of brood and adult bee diseases, including Deformed Wing Virus (DWV), Chronic Bee Paralysis Virus (CBPV) and more, leading to absconding or colony collapse.

Varroa mites prefer to lay their eggs and rear their young in drone brood. The longer development of drones (24 days compared to 21 days for workers) gives the varroa time for an extra brood cycle, allowing the female to lay more eggs and produce more adults.

The mated female varroa mite jumps into the cell just before it is capped and bites into the larva to create a feeding point which allows her to feed on the larva's fat bodies and haemolymph. About 60 hours later she lays her first egg which will hatch within 26 hours and will develop into a male. Every egg laid at 30-hour intervals thereafter will be a female. All the mites feed on the larva's fat body until they leave the brood cell. The male will mate with all the females. The females leave the cell with the emerging adult bee and go on to repeat the cycle, whereas the luckless male dies without leaving the cell.

This has consequences for the adult bees, including the quality of drone semen. Poor sperm can cause reduced fertility in queens, weaker offspring and early queen supersedure in the next generation. In the case of workers, they may have a reduced lifespan, are less able to carry out essential tasks such as raising brood and guarding. Their flying and orientation skills are impaired resulting in reduced income from foraging contributing to winter losses.

The injury to the larva caused by the bite to create the feeding point allows bacteria to accumulate. The mite's saliva stops the wound from healing and permits the entry of damaging viruses such as Deformed Wing Virus (DWV) which is then transmitted into the adult bee population. The emerging bees are smaller, weaker and with shorter life expectancy. Some studies indicate that their immune system may be suppressed.

Parasitic Mite Syndrome

Heavily infested colonies may show signs of **Parasitic Mite Syndrome** which include:

▸ Perforated cappings where the bees have investigated suspected problems.

▸ Dead pupae in their cells.

▸ Partially cannibalised larvae

▸ Varroa openly visible on the adult bees.

▸ Large number of bees with deformed wings.

▸ Increasingly scattered brood pattern.

▸ Chilled or neglected brood where there are insufficient adult bees to feed and keep it warm.

▸ Defensive behaviour.

▸ Dwindling population.

▸ Death of colony.

Damaged brood showing signs of severe mite infestation with neglected brood, damaged cappings and partially cannibalised larvae
Courtesy the Animal and Plant Health Agency (APHA), Crown Copyright

Honey bee with varroa on the lower abdomen where it has access between
the plates to feed on the fat body.
Courtesy the Animal and Plant Health Agency (APHA), Crown Copyright

Varroosis
Deformed and stunted immature bees and multiple varroa on the body of the mature adult

Population growth

A female varroa mite can create 1.5 new mites in a worker brood cell and 2.6 new mites in drone brood. Therefore, the varroa population is doubled every 21 or 24 days.

VARROA MANAGEMENT CONCEPT:

Understand exponential growth —

the more mites in the hive,

the faster they increase!

Mite population	Mites per half cup	Additional mites that month
100	1	100
500	2	500
1,000	5	1,000
2,000	7	2,000
5,000	15	4,000
10,000	35	10,000

Effect of sudden mite invasion (arrow) on subsequent mite growth at different levels of infestation
Image ScientificBeekeeping.com

As can be seen in the graph below, having a low varroa level low at the start of the season has a significant impact on the level of infestation throughout the season.

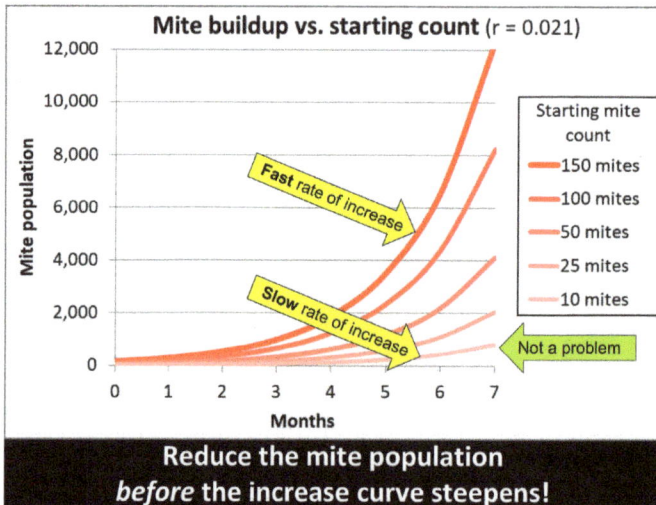

Mite buildup vs. starting count (r = 0.021)

Mite population (y-axis): 0 to 12,000

Months (x-axis): 0 to 7

Starting mite count:
150 mites
100 mites
50 mites
25 mites
10 mites

Fast rate of increase

Slow rate of increase

Not a problem

Reduce the mite population
before the increase curve steepens!

Graph showing exponential growth of varroa population starting at different levels of infestation.
Image ScientificBeekeeping.com

When the population of bees begins to decline in autumn the number of varroa continues to increase, leading to a dangerous rise in the ratio of mites to bees.

Monitoring the varroa load

Regular monitoring helps to keep track of the number of mites in the colony and indicates times when treatment is required.

Monitoring by counting the **mite drop** in hives with open mesh floors is carried out as follows:

▸ Insert the varroa board, marked with a grid pattern to ease counting and coated with a sticky substance or with Vaseline.

▸ Remove the board after a week and count the number of mites.

▸ The varroa drop calculations are made in accordance with the life cycle of the colony and of the varroa mites:

  ▸ Monitoring period 7 days

  ▸ Count the number of mites and divide by 7

  ▸ Multiply the daily drop by one of the following:
    • November to February x 400
    • May to August x 30
    • March, April, September, October x 100

▸ The maximum threshold of mites per colony before serious harm occurs is 1,000.

▸ Treatment should begin before this level is reached.

Sugar roll

This is a non-invasive method which gives an approximate indication of the level of infestation

▶ Take a sample of 400 bees (about 120ml) in a large jar with mesh instead of a lid.

▶ Add 2 tablespoons of icing sugar and roll the jar for a few minutes.

▶ Shake the jar over a white surface to allow the mites to fall.

▶ Count the number of mites and calculate an approximate percentage.

▶ A rate of 3% or more indicates that treatment is required.

▶ The bees can be released in the colony, shaken but not harmed.

Alcohol wash

This method involves the sacrifice of bees:

▶ Take a sample of 400 bees in a jar and add enough 70% Ethanol to cover them.

▶ Leave for 24 hours.

▶ Empty the contents of the jar through a sieve into a white bowl.

▶ Wash the bees again with Ethanol to remove any remaining mites.

▶ Count the mites that fall in the bowl.

▶ 3% or more indicates that treatment is required.

Drone brood uncapping

Uncap about 100 cells and count the number of mites to give a percentage level of infestation.

What we can do about it

By practicing a rigorous programme of integrated pest management, we can keep the level of infestation as low as possible. The management steps include the use of chemical and non-chemical (sometimes known as biotechnical) methods:

▶ Open mesh floors which allow fallen mites to drop to the ground and die rather than climbing back and continuing to live in the brood box.

▶ Drone brood uncapping to remove sections of drone brood and with it a lot of the varroa.

▶ Drone brood removal where whole frames of drone brood are removed and destroyed. This can be done either by adding a frame of drone foundation or by putting in smaller frame such as a super, in which case the bees will build drone cells underneath which can be cut off and destroyed once they are capped.

▶ Queen/Comb trapping is an effective technique for *Varroa* control; however, some time, skill and extra equipment are needed in carrying out the technique. It is claimed to give an efficacy of 95% and one of the advantages is that no chemicals are used when supers are on the hive. For more information see the NBU queen trapping guidance at visit https://shorturl.at/HSU09.

▶ Comb change by shook swarm followed by Oxalic Acid treatment. Some brood is lost but the colony is given a varroa-free fresh start to the season and colonies are often said to enjoy a huge boost in population afterwards.

Drone brood removal using an uncapping fork showing varroa feeding on a larva
Courtesy the Animal and Plant Health Agency (APHA), Crown Copyright

Drone brood uncapped showing varroa feeding on larva
Photo Alan Baxter

Varroa Treatment

Treatment for varroa is applied at different times of the year:

Winter

Oxalic acid treatment is given in winter, usually in December when there is little or no brood at a time when the majority of the mite population is living on the adult bees (often referred to as the Phoretic phase of the varroa life cycle). This can be carried out by trickle, spray, GasVap or sublimation (often called vaporizing or fumigating) in which solid oxalic acid is transferred directly into a gaseous state. Attention must be paid to the manufacturer's instructions regarding precautions during use and the wearing of personal protection equipment.

Spring

Collected swarms which have been placed in isolation are given Oxalic Acid treatment while they're still without brood.

Shook swarms are treated with Oxalic Acid before there is any sealed brood.
Other colonies with capped brood which are found to have a high level of infestation can be treated with a Veterinary Medicines Directorate (VMD) approved miticide. See the table below.

Summer

For strong colonies, late summer treatment with miticides such as Formic Pro (formerly known as Mite Away Quick Strip or MAQS) only requires one dose applied after the honey harvest has been taken. This reduces the number of times the hives are opened. Efficacy in the region of 95% is claimed for this product.

Below is a list of the VMD approved treatments:

Natural or 'soft' chemicals

Name	Ingredients	Form & usage notes
Apiguard gel	Thymol	Gel. Duration of treatment 6-8 weeks
Apilife var	Thymol and essential oils	Strips. Duration of treatment 4 weeks
Api-Bioxal	Oxalic Acid	Powder or solution. Duration of treatment 1 day
Formic Pro	Formic Acid	Strips. Formerly MAQS. Duration of treatment 7-10 days.
Oxuvar	Oxalic Acid	Powder or solution. Duration of treatment immediate
Oxybee	Oxalic Acid, glycerol, sucrose, essential oils	Powder or solution. Duration of treatment Immediate
Thymovar	Thymol	Strips. Duration of treatment 12 weeks
VarroMed	Formic Acid and Oxalic Acid	Liquid. Duration of treatment 3-6 weeks depending on season. Use in winter, spring & autumn.

Synthetic or 'hard' chemicals

Please note that varroa mites can become resistant to all the 'hard' chemicals below. Treatment should be varied to avoid resistance and lower effectiveness.

Name	Ingredients	Form & usage notes
Apistan	Tau Fluvinate	Strips. Duration of treatment 6 weeks
Apitraz	Amitraz	Strips. Duration of treatment 10 weeks
Apivar	Amitraz	Strips. Duration of treatment 6-10 weeks
Bayvarol	Flumethrin	Strips. Duration of treatment 4-6 weeks
Polivar Yellow	Flumethrin	Strips with holes over hive entrance, flying bees collect product when entering and distribute it around the colony. Duration of treatment 4 months minimum.

Please note that all treatments must be recorded on VMD Medical Record form and retained for 5 years. The form can be downloaded from the VMD Website.

Nucs & small colonies

Nucs and smaller hives have fewer bees and less ventilation so need smaller doses. Formic Pro is not suitable for smaller colonies and nucs with less than 6 frames of bees. A thymol-based treatment such as Thymovar may be more suitable. In all cases check the manufacturer's instructions before applying the treatment.

The occasional problems with loss of queens reported by some beekeepers using MAQS/Formic Pro can generally be avoided by having strong, healthy colonies with vigorous young queens, and by following the manufacturer's instructions to the letter, in particular regarding hive ventilation and ambient temperature. I have been using MAQS/Formic Pro for several consecutive years without any problems. If in doubt about the strength of the colony, apply 1 strip for 7-10 days then a second strip for a further similar period.

Unlike MAQS, the Formic Pro manufacturer specifies that it must not be used when supers for extraction of honey are in place.

Insurance

In certain circumstances, paid-up members of the BBKA may qualify for compensation for loss of colonies under an insurance scheme operated by the company Bee Diseases Insurance (BDI). For more information check with your local BKA Membership Secretary.

F2F Strong Colonies

"Superiority of numbers is the most common element in victory."
On War: Carl von Clausewitz

Like most life forms, strength is required to resist pests and diseases. In a honey bee colony this means the fitness of the individual bees and the numerical strength of the colony. Numerically strong colonies are better able to withstand attacks by pests and diseases than weak ones. Before the arrival of Asian Hornets, most of us were used to the population of our colonies increasing gradually from spring onwards to achieve a maximum foraging force in July in time for the main nectar flow.

Unfortunately, that main summer flow may coincide with the start of the Asian Hornet predation period. If the bees can't fly in summer because they are under threat of predation and locked in, they must have achieved maximum income of nectar and pollen before predation starts, and we will have to supplement their food and water stores until the threat has reduced.

Therefore, we need rapid population growth in spring to take advantage of the spring and early summer nectar flows. The problem is that this comes at a time when there is an increasing amount of brood, and the winter bees are coming to the end of their lives. This is why the previous autumn feeding is important to stimulate the queen to produce a lot of strong winter bees with a long life expectancy for the following spring.

Those of us who take colonies to fields of winter-sown Oil Seed Rape will be familiar with the techniques of stimulating rapid spring build up.

Spring Colony Building

The preparation for building strong colonies begins the previous autumn. The aim is to build the colonies to field a maximum possible foraging force in April and May.

Requirements

▸ Productive young queens marked (and clipped).

▸ Healthy colonies with low varroa loads.

▸ Prevention of swarming.

▸ Equal-sized colonies for simplicity of management if there are multiple colonies.

▸ Large number of bees emerging at the end of February/early March to become foragers in April and May.

▸ Brood nest rearranged with sealed brood at the centre and unsealed on the outside to discourage bees from putting honey in the brood box.

▸ Supers, ideally with frames of drawn comb, added once foraging begins.

Potential risks

▸ Swarming.

▸ High ratio of brood to adult bees as the winter bees come to the end of their lives, therefore insufficient adults to care for the new brood.

▸ Workers eating eggs to limit growth of colony if no flow or insufficient bees to feed larvae (worker policing).

▸ Presence of Nosema limiting spring growth.

Table of actions to be taken to achieve strong colonies in spring:

Timing *	Action	Notes
Previous Autumn	Ensure colonies are healthy, strong and well fed going into winter. Feed pollen or pollen substitute if in short supply. Feed thick syrup to top up stores. Introduce new queen if present one is more than 2 years old. Feed thin syrup mid-September to stimulate queen to produce winter bees. Ensure hives have 6 frames of brood and stores and are full of bees. Fit mouse guards and woodpecker protection in October Fit insulation to roof and walls of brood boxes. Secure hives against strong winds with straps or weights. Protect supers or stored frames against wax moth before winter storage.	Varroa treatment applied after honey taken off to protect winter bees. 21kg of stores to remain after honey harvest. Bees born in August/September die in October/November. Bees born in October live until March. If in doubt add frames of capped brood or unite with stronger colonies. Insulated hives are said to increase rates of winter survival and spring growth. Freezing or fumigation with 80% Acetic acid.
Winter	Ensure colonies have enough stores by hefting at least monthly. Oxalic acid treatment early December when little or no brood. If required, until March, stronger colonies, spring growth	Add fondant if required. Trickle, GasVap or sublimation.
End Jan	Start continuous feeding thin sugar syrup, and pollen if needed, to simulate early nectar flow and stimulate queens to lay. Thick syrup will be laid down as stores.	If weather is cold, they may not come up and over onto a rapid feeder so may need to give fondant instead. Contact feeders don't require this. There is a risk that if the feed syrup is stored in the brood box the queen won't have enough room to lay.
End Feb	Spring inspection when temperatures permit. Continue to feed sugar syrup (and pollen if needed).	Check stores, space for queen to lay, change floor, check varroa drop. Treat for varroa if >1,000 per colony.

Early Mar	First full colony inspection. Health inspection. Check queen present & laying. Inspect weekly for signs of swarm preparation. Continue to feed sugar syrup.	Check queen present & laying. Change queen if not performing. Take samples to check for Nosema if build up slow. Swarm prevention measures if required.
Mid Mar	Weekly inspections. Equalize colonies **see below. Inspect weekly for signs of swarm preparation. Continue to feed sugar syrup.	Monitor colony growth. Swarm prevention measures that will not interfere with the honey production if required.
End Mar	Inspect weekly for swarm prevention. Continue to feed sugar syrup.	In theory, if queens are clipped, inspections can be every 10 days, but the last thing you want is for them to swarm, return, but you lose the queen. It's better to carry on with weekly inspections just in case.
Early Apr	Ensure queen has room to lay. Continue to feed sugar syrup. Add supers. Weekly inspections for swarm prevention essential. Swarm control carried out immediately the first charged queen cups are found. Add frame of drone brood.	Brood boxes can rapidly run out of space as the nectar flow is abundant and the colony might swarm. Move full frames to the outside, put empty frames near the centre. Adding an extra brood box gives more space with the bonus of getting some drawn comb, and it gives the growing population of bees something to do. Produce 'good' drones for queen mating and varroa reduction later in the season.

* Timings are for the south of England. Other regions will be later.

** Strong colonies can be held back to reduce congestion, with consequent risk of swarming, and weak ones strengthened, by transferring brood between them providing both are in good health.

Giving brood alone to a weak colony may not work if there are too few young nurse bees to look after it, so proceed as follows:

- Place a super of drawn comb or foundation on the strong colony, without a queen excluder, and allow the queen to lay in it.

- When the super is full of brood and bees, smoke the bees and the queen down into the brood box, ensuring that there is enough room for the queen to continue to lay.

- Once the brood is capped, put on the queen excluder and check after a few days that there are no new eggs in the super and therefore the queen is in the brood box.

- Put the super on the weak colony which will soon have a boost in population of young bees.

Well-insulated hives

If hives are well-insulated, the colony will have used fewer resources and less energy keeping warm through winter, the internal temperature of the brood nest will be higher, and the queen may start laying earlier.

The winter bees will be more rested, have more resources left in their fat bodies and therefore be better able to look after spring brood.

See reference section for work by Derek Mitchell of Leeds University on the benefits of all-year-round insulation.

However, some commentators point out that, in areas that enjoy mild winters, insulation may allow queens to continue laying all winter. This means that brood break that is useful in interrupting the varroa breeding cycle and reducing the number of mites, doesn't occur.

Early Spring Feeding

Feeding thin sugar syrup and pollen supplement in very early spring stimulates the queen to start or increase laying.

In the case of honey-producing colonies, feeding should stop as soon as the spring flow starts and supers are added to avoid contamination of the honey with sugar syrup.

Rigorous Swarm Prevention and Control

As part of the Fit2Fight programme we are asking our queens to produce a lot of new bees in the early part of the year.

Swarming is a serious risk during rapid spring build-up, so it is vital to understand and recognize the early signs of swarm preparation, to watch them like a hawk, and take prompt swarm prevention measures as described below.

With the threat of Asian Hornets hawking in our apiaries in a few short weeks' time, we cannot afford to lose any bees. If we do, there is less chance of building numbers up for the spring flow and going into the predation period with a big, strong colony.

In order to deal with the ever-present threat of losing half your bees and disturbing your neighbours, it is useful to understand the swarming process, the key indicators to look out for, and the prevention or control measures you can take.

Why do honey bees swarm?

Swarming is the natural way that colonies use to reproduce themselves. It happens mainly in May, June and July, but can continue into September.

There are a number of conditions, or a combination thereof, that are likely to trigger the swarm impulse including:

▶ Older queen. Her glands are producing insufficient quantities of the pheromones known collectively as queen substance, which has the effect of inhibiting the creation of queen cells.

▶ Overcrowding. The queen's pheromones aren't reaching all parts of the hive, the queen has no room left to lay eggs, young bees are secreting wax with no space to put it and the receiving bees have nowhere to store incoming nectar and pollen brought in by the foragers.

▶ The queen is producing more brood and the ratio of house-bound young bees to older foragers is out of balance.

▶ Fine weather. A spell of warm, dry spring or summer weather is ideal for bees to leave the warmth and security of their home to find and furnish a new one.

▶ Nectar flow. A strong flow means that the new colony will have a plentiful supply of food required to build comb and to feed the larvae.

▶ Spell of confinement. After a cold or wet spell, the stir-crazy bees may swarm on the first fine day!

In the countdown to swarming chart below you will see the progression that occurs. I have divided it into two parts: the times when you can take **prevention** measures: the times when **control** is necessary.

Swarm Prevention

Prevention requires regular, thorough inspections and careful observation weekly.
Here are some simple but effective measures you can take:

▶ **Young queens**

- Queens are less likely to swarm in the first full year of their lives so replace them at least once every 2 years.
- Older queens with good characteristics, or that you can't bear to lose, can be 'retired' to nucs and used for raising more queens.

▶ **Mark your queens**

- If you cannot find the queen, get another pair of eyes or two to help you.
- The earlier in the season the better when there are not too many bees.
- Do it in the warmest part of the day when the maximum number of bees are out foraging.
- Or move the brood box to one side and put a super on its original stand to divert all the flying bees and give you more room to look.
- The queen is more likely to be in the brood area, but don't forget to look inside the brood box in case she's fallen off.

▶ **Clip your queens**

- This involves gently cutting off the end third of one wing.
- It is not to everyone's taste but there are no nerves or blood vessels there – it is like cutting your toenails.
- Practice a few times with drones to give you confidence.
- If you are struggling with any of these, get a more experienced beekeeper to help you. People are always happy to lend a hand.

▶ **Add space**

- The bees need room to live and to store incoming nectar and pollen.
- Nectar contains about 80% water and requires a greater volume of space than honey which only contains less than 20%.
- The queen needs space to lay her eggs.
- Add a super of drawn comb as soon as the brood box starts to look crowded, or the bees start to build comb on top of the brood frames. Frames of foundation are not

seen by the bees as space and won't have the desired effect.
- When the flow starts, they will be bringing in nectar very fast so be ready to add more supers as soon as the current one is 2/3 full.
- If there are not enough supers they will store nectar in the brood box, thereby depriving the queen of laying space and she will stop laying or they might swarm.
- Remember that during the day, a lot of the bees will be out foraging. At night they need somewhere to go.
- During a flow, replace capped frames of honey with drawn comb or foundation. Drawing comb keeps them busy and out of mischief.
- Replace damaged comb which cannot be used efficiently.
- Be prepared by having spare equipment ready in advance to take off the pressure when the time comes to act.

Swarm Control

Once queen cells have been formed, the time for prevention has passed and control measures are needed. An artificial swarm is a way of tricking the bees into thinking they have swarmed by separating the queen and flying bees from the brood and the nurse bees. It is called 'artificial' because in a natural swarm the queen will be accompanied by bees of all ages, not just the older flying bees.

I prefer to use either the **Nucleus** or the **Pagden** method if I want to make increase from the colony, or the **Demaree** technique if I don't want any more hives or to breed from that particular queen.

For details on these methods of swarm control please see Annex F, G & H.

Beware the cast, or secondary swarm
After the first swarm has issued, and the old queen with up to 70% of the bees of all ages has left the colony, the first of the new queens will emerge from the queen cells left behind. She will attempt to eliminate the competition by fighting with any others and stinging the sides of the queen cells. The sealed brood that the queen had been producing beforehand will start to emerge (remember those big slabs of brood about 3 weeks ago) and they will be ready to swarm again. To reduce the risk of losing any more bees:

▸ Remove all the sealed queen cells.

▸ Choose the best unsealed one, marking its position on the frame with a drawing pin.

- A few days later go back and remove any more queen cells that have been made. Don't do this in the middle part of the day to avoid the chamce of disrupting the new queen on her mating flights.

- Leave alone for three weeks to allow the new queen to mate and start laying eggs.

- Make notes in your hive records of the dates – it can take longer than you think for a new queen to be mated and start to lay.

Don't be beguiled into thinking that your colony hasn't swarmed because the hive is still full of bees. The queen's rate of lay increases at the start of swarm preparation so it could be all that capped brood which has emerged after the main swarm has left!

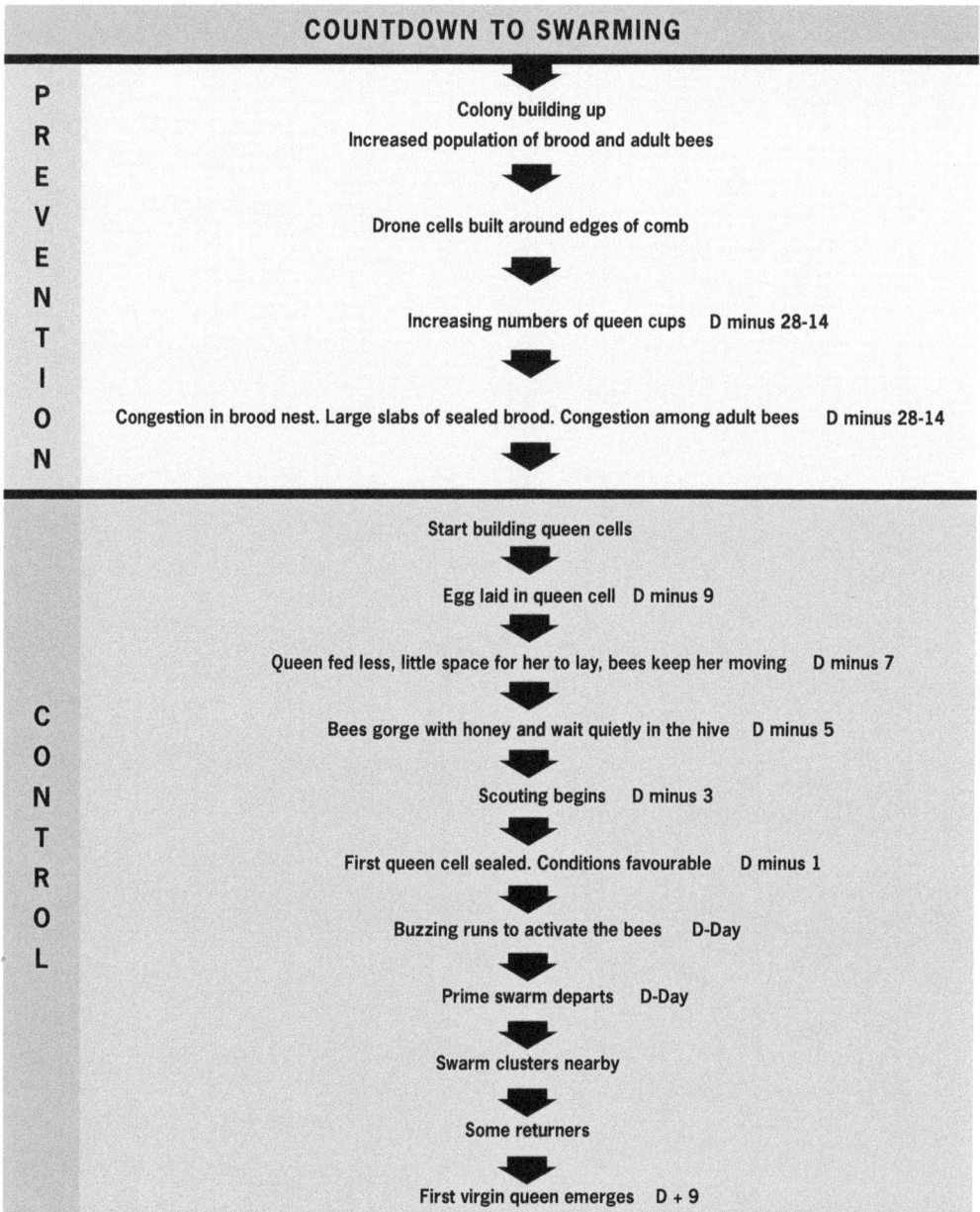

COUNTDOWN TO SWARMING

P R E V E N T I O N

Colony building up
Increased population of brood and adult bees

Drone cells built around edges of comb

Increasing numbers of queen cups   D minus 28-14

Congestion in brood nest. Large slabs of sealed brood. Congestion among adult bees   D minus 28-14

C O N T R O L

Start building queen cells

Egg laid in queen cell   D minus 9

Queen fed less, little space for her to lay, bees keep her moving   D minus 7

Bees gorge with honey and wait quietly in the hive   D minus 5

Scouting begins   D minus 3

First queen cell sealed. Conditions favourable   D minus 1

Buzzing runs to activate the bees   D-Day

Prime swarm departs   D-Day

Swarm clusters nearby

Some returners

First virgin queen emerges   D + 9

Countdown to swarming
Original author unknown, chart modified by Alan Baxter

Weak colonies

One strong colony normally has more than double the chance of surviving predation than two weak ones.

Uniting colonies in spring enables them to field a bigger foraging force for the spring flow and go into predation lockdown with lots of bees armed with plenty of stores. However, this may stimulate swarming which reduces colony size so careful attention should be paid to swarm prevention as described above. The newspaper method is the simplest and most commonly used.

Colonies that are struggling to increase numbers during May should be united in plenty of time before predation starts as long as they are in good health.

It is worth asking yourself why they are weak at this time of the year. For example:

▶ Is the queen young and normally productive ?

▶ Has the queen been superseded? If all your queens are marked you will know.

▶ Is it a problem with the queen? Does she need to be replaced?

▶ Have they had enough food ?

▶ Was the build-up normal in the spring?

▶ When did you last do a full health inspection? Was everything normal?

▶ Might they have a disease such as Nosema which is preventing them from growing? If in doubt arrange for a sample of bees to be tested by microscopic examination. Local or Area Beekeeper Associations may be able to advise on where this can be done.

▶ Could they have swarmed and you didn't notice?

▶ When was the last varroa count? Do you need to treat them now rather than wait until July?

Young productive queens

To produce the colonies best suited to the task of fighting the Asian Hornet, ideally queens will have been selected for:

▸ Rapid spring buildup.

▸ Low swarming characteristics.

▸ Hygienic behaviour for disease resistance.

Whilst selecting and raising queens might not be feasible for many hobbyists with small apiaries and limited time, Local BKAs, or groups of individuals, can often help by supplying suitable queens.

However, if you need to buy in queens, those raised in other areas will adapt less well to your local conditions. It's much better to find one from a local breeder or from your Beekeepers' Association. See NBU and BBKA advice regarding importation of queens.

Queen rearing timing

Queen rearing needs to be concentrated into spring and early summer and finished before predation starts. Giving colonies drone foundation in early spring has a triple benefit of:

▸ Stimulating queens to produce drones for early queen rearing.

▸ Improving the population of the local drone congregation area with your 'good' drones.

▸ De-capping or removal of drone brood to reduce varroa as part of Integrated Pest Management once queen rearing is finished.

Mating nucs are extremely vulnerable to predation because of the small number of bees so new queens should be introduced to full colonies, or 14 x 12 nucs as soon as possible after they start laying and before predation starts.

Overwintering nucs

Nucs, preferably 14 x 12, with 6 frames, for the overwintering of new queens should be bursting with bees by the time predation starts. Group them together for extra security (see section on defence in the apiary).

Strong colonies going into winter

"In the life of the honey bee colony there are only two seasons –
winter and preparing for winter"
André, Maître Apiculteur, Bellevigne-en-Layon

To ensure that colonies survive the winter and have enough adult bees to support the spring surge in brood population, a strong healthy population of winter bees needs:

▸ A productive young queen.

▸ Well- developed fat bodies.

▸ Good health and a low varroa load.

▸ Access to plenty of stores.

Giving colonies syrup and pollen throughout the period of predation, and into autumn, stimulates queens to continue laying in late summer and early autumn, and mid-September and early October to produce good winter bees. Bees emerging in August/September will die in October/November whereas bees emerging after the beginning of October are likely to survive until the following spring.

Pests and predators

Wasps, mice, wax moth and woodpeckers are the usual suspects in autumn and winter.

In 'waspy' years these normally beneficial insects can be a nuisance. A strong colony will normally deal with wasps, but weaker ones can be vulnerable. Wasps are known to chew holes in unpainted poly nucs or brood boxes to access the stores inside:

Wasps chewing into a poly nuc.
Photo by Alan Baxter

Simple non-kill wasp traps can be deployed in this short, early autumn period but should be checked daily to release any innocent bycatch such as European Hornets. See the chapter on Traps and Trapping for more information about designs, bait, killing the targets and releasing non target species.

Mice are sometimes tempted into hives at the start of winter by the prospect of a warm home in which to hibernate. Mouse guards fitted in October will deny them access.

Empty super frames are put in the freezer for at least 48 hours to kill the wax moth and Braula coeca eggs, larvae and pupae.

In a hard winter when there is little food available, green woodpeckers will drill holes in the side of hives and raid the contents causing considerable damage and often leaving the colony to die of cold or starvation. A cage of chicken wire around the hives will prevent the birds from attacking them.

Insulation

A layer of insulation in the roof and on the sides of wooden hives can help to keep the cluster warm and reduce the quantity of stores they consume to keep warm. The winter bees are likely to be more rested and to live longer. Insulated hives are claimed to enjoy better winter survival rates and faster spring build up. Poly hives and nucs are said to have an advantage in this respect.

Summary

- A colony that is fit, strong and well fed, with a productive young queen, will have a better chance of getting through winter and spring.

- Bees born in August and September will die in October and November. Bees born in October will live until the following spring.

- Feeding of **thin 1:1** sugar syrup in mid to late September will stimulate the queen to lay the eggs that will emerge in October as the vitally important winter bees.

- Feeding of **thick** sugar syrup to be laid down as stores.

- Feeding with fondant will see the colony through times when it is too cold for the bees to consume syrup.

▶ Varroa treatments in August and December are required to prevent winter and spring losses.

▶ Colonies led by young queens are much more likely to survive than those with older queens. They are also less susceptible to swarming the following spring.

▶ Protection against mice, and woodpeckers is fitted in October.

▶ Insulation can make stores last longer, extend the life of the winter bees, and help the colony to grow faster in spring. See earlier remarks about insulation in the areas that enjoy mild winters.

For more on winter preparation see Annex H.

F2F - Well-Fed Stock

Drawing by Toby Melville-Brown

"An army marches on its stomach"
Frederick the Great, King of Prussia 1740-1786

As with a successful army, good nutrition is an essential part of any beekeeping operation. It is even more important if the colonies are under threat from Asian Hornets and at times when they cannot fend for themselves.

This section covers different feeding requirements throughout the year.

Sugar syrup and fondant

Sugar in the form of nectar is an essential component of the diet of adult bees and larvae. It provides energy for internal bodily functions, for work inside and outside the hive, maintaining the temperature of the brood nest all year round, for generating heat when making wax to build comb, and more.

Spring We have already seen that we need to build up the colony early to have a big foraging force for the spring flow. We feed thin syrup to simulate a nectar flow which encourages the queen to increase her laying rate and, after comb change, to stimulate the bees' wax glands for comb building.

Syrup can be given if daytime temperatures are regularly above 10°C. A 'thick' solution of 2:1 syrup allows for more efficient storage, a 'thin' solution of 1:1 may be consumed without dilution and therefore useful for immediate use in wax building, feeding larvae and in stimulating the queen to lay.

Summer and autumn during the period of predation Syrup is given continuously to replace loss due to reduced foraging, to preserve or top up their existing stores, to generate water by evaporation (see the section on water below), to build up stores and for queens to produce strong winter bees with well-developed fat bodies.

In mid to late September and early October, stimulative feeding of thin sugar syrup, and pollen substitute if pollen isn't being brought in, will encourage queens to continue laying at a time when they normally wind down in preparation for winter. This boost in population of winter bees helps to reduce winter losses and ensures enough workers to look after the queen and brood the following spring.

The proportions and uses of syrup are as follows:

Concentration	Sugar	Water	Use by bees
Thick	1 kg	625 ml	Laid down as stores. Requires dilution before consumption.
Thin	1 kg	1250 ml	Immediate consumption. Stimulating the queen to lay, workers to produce wax and draw comb, feeding larvae.

Types of feeder

There are several types of feeder in current use:

Contact feeders in the form of a tub or bucket with 1.5 mm drilled feed holes or a mesh-covered opening in the lid. When filled with syrup and inverted over a drip tray, a partial vacuum is formed. It is placed over the hole in the crown board allowing the bees to feed from the opening. An eke is required to create enough space for the roof to fit securely. Despite being sometimes described as 'rapid' feeders, the rate of delivery of the feed is relatively slow, more closely resembling that at which nectar arrives during an autumn flow; this is said to encourage the queen to lay. Sizes range from 1 Litre to 4.5 Litres capacity.

Contact feeder

Tray feeders or 'rapid' feeders consist of a wooden or plastic tray with a trough from which the bees can feed without drowning. Round ones are placed over the hole in the crown board and need an eke to create space. Square ones are the same dimensions as a brood box and sit directly over the frames with a crown board on top. The rate of delivery is fast, and the syrup is more likely to be laid down as stores.   Popular designs are:

English feeder

Ashforth feeder

Frame feeders are hollow frames made of wood or plastic, the same width as a brood frame, with lugs to hold it in place in the brood box. A float or a tangle of dried plant material allow the bees to feed without drowning.

Frame feeder

Winter Fondant is given to maintain a reserve supply of food. Supplied in the form of a 12.5 kg block or in handy 1 kg sachets, it is available from most beekeepers' suppliers. Commercially sold baker's fondant can contain additives which are unsuitable for feeding to bees, so be sure to read the contents on the label first. The fondant needs to be diluted to 50% water before the bees can eat it; they usually use condensation in the hive for this.

Block of fondant

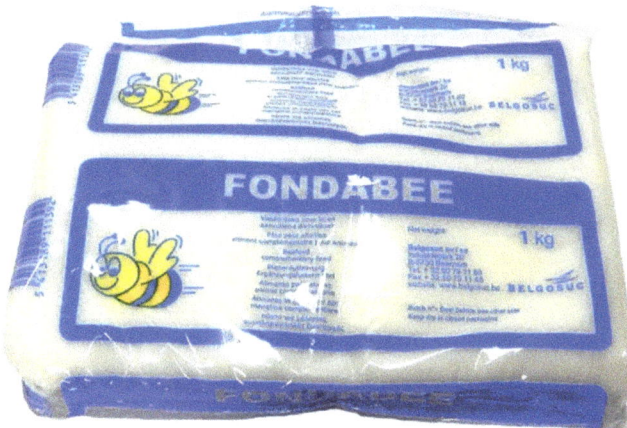

Sachet of fondant

Note: White sugar does not contain the trace elements, amino and fatty acids, vitamins and live gut bacteria that are found in nectar and pollen. It is infinitely better to leave the bees with enough of their own stores in the form of honey, particularly if they have been subsisting on sugar syrup during lockdown.

Honey bee foraging for nectar.

Pollen

Pollen is a source of protein and lipids used by bees for the following:

▶ To stimulate the queen to lay eggs.

▶ To develop young adult bees' hypopharyngeal and mandibular glands. These glands produce the vital elements for brood food and royal jelly in young bees on larva feeding duties, and the enzymes Sucrase, Diastase and Glucose Oxidase in older bees when processing nectar into honey.

▶ To stimulate the development of the wax glands in slightly older bees and provide the raw material for making wax when building comb.

▶ To provide protein and lipid content in brood food and royal jelly.

▶ To mix with nectar when making bee bread for storage.

- ▸ To build up fat bodies which sustain bees in winter and enable older bees to feed spring larvae.

- ▸ For growth and repair of tissue.

Pollen is to be found in flowers most of the year in Britain. Some early flowering varieties such as hazel, alder and willow provide an early source. Other flowering plants including Ivy, Ragwort and Himalayan Balsam are plentiful until late autumn.

The nutritional value of different pollens is varied. Somehow, the bees know this and will often be seen ignoring a nearby source in preference to a better one much further away.

In all cases the ability to forage is dependent on the weather and there are huge regional variations in the timing of the seasons and the prevailing weather conditions.

Honey bee foraging for pollen.

Pollen is fed:

Early Spring - to stimulate the queen to start or increase laying for rapid colony build up.

During predation in late summer and autumn - to replace loss from reduced foraging and to keep the queen laying.

September/October - for queens to produce strong winter bees with well-developed fat bodies.

It can be in the following forms:

Pollen **supplement** which is a mixture of ingredients fortified with natural pollen.

Pollen **substitute** consisting of ingredients that replace natural pollen but lack many of the nutrients that natural pollen provides.

The constituent parts of each of these are:

Ingredient	Pollen substitute %	Pollen supplement %
Fat-free Soya flour*	75	60
Brewer's yeast	25	20
Natural pollen	0	20
Total	100	100

*Dried fat-free skimmed milk and dried egg yolk can also be used.

Method

Mix the ingredients with sugar syrup or honey into a thick paste to form convenient sized patties about ½" thick and spread onto waxed paper. Lay the patties flat on top of the brood frames with the wax paper uppermost.

For many of us, collecting pollen and making our own pollen patties is too much faff. There are plenty of commercial products available, but care should be taken to read the list of ingredients carefully to ensure that the pollen, or pollen substitute, content is high enough to give any benefit.

CANDIPOLLINE GOLD

Sachet of Candipolline

Note: Research at the University of Florida has thrown doubt on the ability of bees to digest, and pass on to larvae, many of the ingredients in pollen substitutes *(Noordyke, Ellis et al)*.

Water

Water is used by bees as follows:

▶ Homeostasis including maintaining the fluid balance level of adult and brood haemolymph.

▶ Diluting stored honey to 50%, at which point it is able to be consumed by the bees.

▶ Controlling the temperature of the hive by evaporation.

Bees obtain water from evaporation of nectar when it's being processed into honey, and from foraging. There are no means of storing water in the hive, but water foraging or 'reservoir' bees can temporarily store it in their own bodies. If the bees are unable to go out foraging, we must provide a source of water for them.

A simple rapid feeder over the hole in the crown board with an eke to give it space is all that is required.

Honey bee foraging for water.

Hot weather

Bees use water to cool the inside of the hive by evaporation. They spread droplets of water on the surface of the comb and fan it with their wings.

Other bees fan air in and out through the entrance. Carbon dioxide is removed and replaced with fresh air at the same time.

In addition, bees cluster on the outside of the hive, behaviour called bearding, to reduce crowding inside and improve ventilation.

During predation the normal temperature regulation system of the hive is disrupted. There is no foraging for nectar or water, it's too dangerous to beard outside in daylight, hive entrances are reduced, and ventilation is restricted.

We can help the bees to maintain the correct hive temperature, humidity and CO_2 levels by:

▶ Providing shade in the hottest part of the day.

▶ Opening entrances in the evening when hornets have stopped flying to allow cooler, fresh night air to circulate, and for undertaker bees to remove the dead and other debris.

▶ Feeding water to compensate for lack of water income from foraging or evaporation of nectar.

Note *Opening hives for feeding must be carried out in early morning or evening when Asian Hornets are not flying.*

F2F Defence

Preparing the Battleground

"Ground is the handmaid of victory"
Sun Tzu: the Art of War

Integrated management in the apiary mainly employs measures which are standard good beekeeping practice, but at a higher level of intensity or frequency. Extra equipment for the protection of hive entrances and to eliminate Asian Hornets hawking in the apiary may be required.

In this section we'll look at:

▶ Apiary biosecurity.

▶ Reducing temptation.

▶ Physical defences.

Apiary biosecurity

Discourage other beekeepers from visiting unless necessary, for example if you need help from your bee buddy, mentor, or a Bee Inspector. If you do expect guests, check their clothing and equipment are clean, give them disposable gloves, and ensure that they apply the same rules of apiary hygiene as you. See the chapter on apiary hygiene for more information about protecting the bees from infection or infestation.

Reducing temptation

Asian Hornets are attracted by olfactory signals (smells). An open hive releases lots of deliciously inviting aromas of pollen, honey and brood so it's a clear invitation to prowling hornets looking for food. There is evidence that hornets are also attracted by the smell of geraniol, a component of the bees' Nasanov pheromone used to gather returning bees at the hive entrance.

Only open the hives for a specific purpose and do quick in and out visits where possible. You don't always need to see the queen.

Two exceptions are for swarm prevention and health inspections when the frames need to be shaken free of bees and examined very carefully.

Carry out all manipulations that involve opening hives before the predation period starts.

If you do need to open the hives for any reason, it should be done early in the morning or in the evening as Asian Hornets don't fly at night.

Physical defences

The level of defence will depend on the traffic light early warning system indicating different levels of threat:

Green: I'm in an area where no hornet nests or sightings have been reported yet.

‣ Grow grass round the hives or put a skirt of boards to stop the hornets getting under and behind the hive, and to give the bees somewhere to hide.

‣ Don't attract attention to the apiary by opening hives or feeding them during the day.

‣ Group hives together to provide safety in numbers. Colours or shapes painted on the front of the hives help orientation and reduce the risk of drifting.

Amber: Asian Hornets confirmed in my area.

▶ Insert Varroa boards in hives with open mesh floors to reduce the stress on the bees from hornets flying under the floor.

▶ Place a woven tangle of twigs and branches in front of hives to keep the hornets at least 1 metre from the entrance, to give flying bees more dispersal space and somewhere to hide. The gaps should be too small for the hornets to fly through but big enough for the bees. This has a similar effect to a muzzle but costs nothing. Hawthorn and blackthorn, pinned to the ground with sticks or pieces of bamboo to stop them blowing away, are ideal for this.

▶ Instead of branches fit muzzles and put out electric harps if you have them. By doing this early the bees will have time to learn their way in and out. For more on defence equipment see 7.4 below.

▶ Single hives are particularly vulnerable to attack so if you're a one-colony beekeeper you might consider moving your hive to another apiary where it can benefit from being with others. If you do this, pay careful attention to the health and varroa load of your own hive and those of the apiary to which you're moving.

▶ Prepare to take off your honey crop if there is one.

▶ Get ready to apply varroa treatment.

Red: Asian Hornets have been seen actively hunting in or near my apiary.

▶ First of all – don't panic. You and your bees are already well-prepared for this, and you are in charge, not the hornets.

▶ Take photos and report the incursion on the AH Watch App to inform the National Bee Unit (NBU) and your nearest AH Coordinator.

▶ Reduce entrances to 5.5 mm.

▶ Prepare to put out 'decoy' traps, but to avoid attracting hornets don't deploy them unless predation is actually happening.

▶ Prepare to start feeding to compensate for reduced foraging (see the chapter on well-fed stock).

- Provide shade and water in hot weather as the colony's normal temperature regulation mechanisms will be disrupted.

- If predation is heavy, think about moving your hives to another site if possible.

- Stand by to assist the NBU or Asian Hornet Action Team (AHAT) in tracking the nest if required (see Chapter 9 on tracking and bait stations below).

- Have fun swatting hornets with your badminton racquet. Asian Hornets aren't usually aggressive away from the nest but beware of being too casual and wearing PPE is a wise precaution!

Defence equipment

In the area of France where I lived the practical, thrifty country folk are adept at make do and mend, using naturally available material they can adapt or reuse. In the case of the Asian Hornet, they devised a system of a woven mesh of twigs and branches to make a screen in front of the hives to protect the entrances from predating hornets. These proved to be highly effective in reducing pressure on the colonies and are easy to make. They require access to a supply of untrimmed banks or hedges such as bramble, blackthorn and hawthorn which may be less readily available to urban dwellers.

Whilst devices including the muzzle and electric harp have been developed in recent years, and are in widespread use, many of the rural beekeepers in my former region continue to use brushwood.

Entrance reducers

Entrance reducers make it easier for the hives to defend the colony with fewer bees. Reducing the entrance to a maximum of 5.5mm is the most ideal, denying entry to the Asian Hornet and helping to reduce the stress of predation.

In addition to the standard entrance blockers supplied with most hive floors, the following are among the many commercially produced devices that have proved effective:

HiveGate is designed to make the entrance easier to defend by extending the colony entrance to reach the centre of the bee cluster. Originally intended to defend the hive against robbing by wasps and other honey bees it is equally applicable to defence against the Asian Hornet.

The Muzzle and the **Electric Harp** have been trialled and used extensively in France, Spain and other European countries, proving to be highly successful in reducing stress on the bees.

La Muselière– the Muzzle
Supplied by E.H.Thorne Ltd

Muzzle with netting

"Stop-it"
Supplied by E.H.Thorne Ltd

Muzzles provide a protective screen and a safe place for bees to hide in front of the hive. The size of the hole is optimised to allow the bees to fly in and out, without knocking off pollen loads, but restricts access for the hornet. The muzzle keeps the hornets at a distance from the hive entrance, provides somewhere for the foraging bees to hide. It also acts as an extension of the colony's defensive territory into which the hornets are reluctant to trespass. DIY models are relatively simple to make. The optimum mesh size has been determined as

95

25 mm, small enough for the bees to fly straight through but big enough to deter the hornets and disrupt their hawking behaviour. A reduction of over 40% in foraging paralysis has been reported when muzzles or brushwood defences are in place.

La harpe éléctrique, arpa electrica, the electric harp

Powered by solar, battery or mains, the electric harp has been found to reduce the pressure of predation on hives, thereby reducing foraging paralysis by a significant amount.

The harps are placed between the hives or at right angles in front of the hives to capture and kill hornets as they circle the colony searching for prey. The gaps between the electrified wires are spaced for the bees to fly safely through, but for Asian Hornets to be electrocuted as they touch the wires with their wings or drowned in the collecting tray underneath. Vast numbers of hornets can be killed by this method and the bycatch is reported to be as low as 10%.

Apiaries in which harps are deployed report a significant reduction increase in foraging with higher levels of brood production and winter survival rates.

F2F Counter Attack

Salting the Battlefield

When the Roman Army was victorious, they would throw salt on the battlefield so that the land became useless and the conquered, as a people, were utterly destroyed.

Roman frieze depicting the Roman Army in battle.

We cannot expect to utterly destroy the Asian Hornet, but we can reduce the impact that it has on our apiaries. This is how we take the fight to them once they arrive in our apiaries.

Trappin' and zappin'

The subject of trapping is controversial, mainly concerning its effectiveness and its impact on other species. It can risk pulling beekeepers in different directions as new thinking continues to emerge and new solutions develop. The suggestions that follow are based on the experiences of those endeavouring to manage Asian Hornets on the Continent and the Channel Islands. Ideas will continue to change as the threat increases in the UK.

Naturally, we all want to be fully pro-active, use the best equipment and techniques available, and protect our bees.

However, traps are not a silver bullet that will solve the problem of the Asian Hornet in our apiaries – they are just one weapon in our defensive armoury as part of an Integrated Apiary Management Strategy.

To protect biodiversity, we must avoid catching insects of other species as far as possible. Even if they are released, the experience of being captured and imprisoned with other insects is highly stressful. It can have serious effects from which they may not fully recover. They may not stay alive for very long after release or be able to reproduce. Traps left out at night will continue to catch other insects after the Asian Hornets have stopped flying for the day.

Whatever the manufacturers claim, as far as we are aware, no trap has yet been produced that is 100% guaranteed to avoid catching innocent victims.

The responsible, measured and proportionate use of traps, with the care necessary to limit by-catch, can be an effective way of slowing down the spread of the invasion and reducing the level of predation in our apiaries.

Effective selective trapping and the location and destruction of nests are vitally important elements in reducing the spread of the Asian Hornet and helping to mitigate its impact in the area where is has become established.

By-catch in non-selective trap
le Museum national de l'histoire naturelle

Wick pots like this one are safe for other species:
Courtesy Marc Struye
European Hornet and butterfly on wick pot

The following guidelines have been issued by the British Beekeepers Association:

GUIDANCE FOR

ASIAN HORNET MONITORING AND TRAPPING

HOW TO MONITOR IN YOUR AREA

Monitoring is vital to stop Asian hornets *(Vespa velutina)*. Nests must be found and destroyed to prevent establishment. **Report any sightings.** Communicate with your local Asian Hornet Team to identify high risk areas.

RED ZONE

Within a 5km radius from where a nest has been found in the last year, and any port or place deemed at high risk.

Use **selective traps**, 1 per 1km² throughout March-November.

Only trap within an apiary if predation occurs, to avoid attracting hornets.

AMBER ZONE

Within a 5-10km radius from where a nest or hornet has been found in the previous year.

Use **bait stations** 1 per 1km² throughout March-November. Placed in an easily observed location.

Selective traps can be used if checked regularly.

GREEN ZONE

Beyond 10km, and where no hornets or nests have been found in the previous year.

Use **bait stations** throughout March-November. Placed in an easily observed location.

Adjust apiary management to improve protection and promote healthy bees.

AVOID BYCATCH

No trap is 100% guaranteed to avoid bycatch. Check all traps regularly.

Even if released, the experience of being trapped is highly stressful for insects.

Relocate your trap If it attracts lots of native insects.

Reducing native insect populations could increase predation on honeybees.

BEST PRACTICE TO AVOID BYCATCH

There is no bait that will attract only the Asian hornet *(Vespa velutina)*, so it is recommended to always:

- **Avoid using kill traps.** Use selective traps or open bait stations
- **Check daily** and release all bycatch
- Avoid filling containers with liquid, instead soak sponges with bait or use a cloth wick

TYPES OF MONITORING EQUIPMENT

A wide range of trapping and monitoring equipment is available to buy or make, with different intentions and results (small sample shown below).

Bait Stations	Selective Traps	Homemade Traps	Kill Traps
Chloe Underwood	Barry Duke	Tony Warren	Chloe Underwood

LOW BYCATCH ←——→ HIGH BYCATCH

Terri Burt	Ruth Green	Terri Burt	Ruth Green
Allow monitoring with no bycatch	Selective cone entrance with open mesh sides	Effective if selective designs are used	Even with adaptations bycatch is high

**More information including the risk assessment and trapping protocol is available at www.bbka.org.uk/resources-for-asian-hornet-teams/
SCAN THE QR CODE**

THE BRITISH BEEKEEPERS ASSOCIATION · FOUNDED 1874

BBKA Monitoring and Trapping Guidance

Types of trapping

Baited equipment is currently used in 5 different ways, taking place at different times of the year. The objectives are different, the logistics are different, the execution is different, the outcomes are different, and it's important not to confuse them. They are:

▸ Spring queen trapping.

▸ Monitoring.

▸ Bait stations.

▸ 'Decoy' trapping in the apiary during predation.

▸ Autumn trapping.

Types of trap, bait, record-keeping and disposal of trap contents are also covered in this section.

Spring Queen Trapping

Background

During the previous Autumn, the majority of Asian Hornets in an original parent nest will have died while large numbers of newly-mated queens (known as gynes) left to find their own spots for hibernation during the winter, often within about 200 metres from the parent nest.

After the rigours of the winter months, the surviving foundress queens emerge from hibernation anytime from late February to late May. These queens are each looking for a secure, sheltered place to build their first embryo or primary nest near sources of food and water located within an average of 700 metres from the parent nest.

It takes 2 to 3 weeks for the emerging queen to build up her strength and for her ovaries to develop. She then begins the search for a suitable place to build her first nest.

The emerging queens are intensely competitive, fighting each other to the death for the occupation of nests, with high attrition rates. A queen emerging from hibernation tries to evict another queen from her embryo nest rather than building her own and takes a gamble on winning the ensuing battle. However, both can be injured in the struggle and risk that neither goes on to found a successful colony. The eviction from the nest of the original queen and her replacement by another is known as usurpation.

If she survives this far, each young foundress queen will be weakened and hungry, and must feed herself before she can lay her first eggs. As head of a single parent family, she has sole responsibility for going out to find food for herself and her offspring. She can't travel far, or the eggs and larvae will die of cold. She must feed and return to the nest as quickly as possible. Hence, although young queens have been known to forage up to 1km from the primary nest, in practice a radius of 600 metres has found to be more likely.

Queens will continue to forage and be vulnerable until the end of May when the first workers emerge and take over the role of foraging and nest expansion, leaving the queen to lay eggs for the rest of her life.

A very small proportion of new queens will fly, be blown, or be transported further afield, in which case they will escape a local Spring trapping campaign.

Traps are placed **in an area where active nests were located the previous late September, October or November, or not found until later in the winter**.

Spring trapping can protect colonies whose survival was threatened by Asian Hornets the previous year by targeting places where nests were present and those affected apiaries.

Recently, the French National Plan moved from a form of widespread saturation trapping towards a more localised, intensely targeted approach.

Threat level: amber and red

Aim

To prevent emerging queens from setting up new colonies.

Method

- ▶ Traps can be put out in February when daytime temperatures average 12°C for 4 or 5 consecutive days and removed at the end of May.

- ▶ An evenly spaced network of traps is placed around last year's nest or the apiary.

- ▶ There are normally up to 10 traps distributed within a radius of 500 m.

- The ideal distance between the traps is no more than 350 metres.

- The grid structure provides coverage but may be adjusted to target sources of forage and water.

- Traps are set up in the sun facing south or south-east in an area around the old nest or the apiary as shown below:

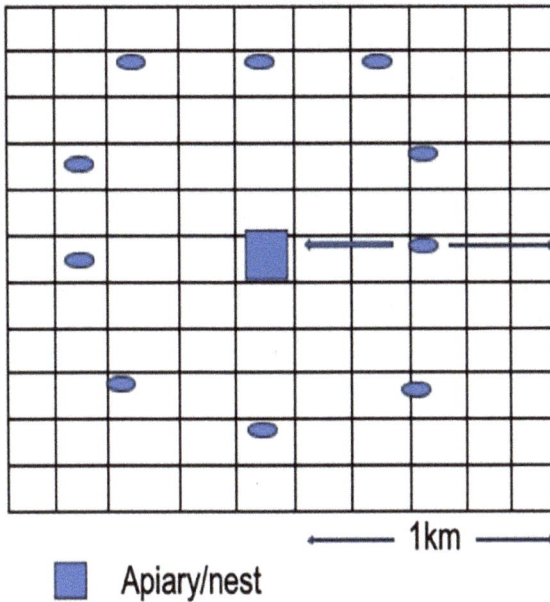

█ Apiary/nest

Layout for spring queen trapping

Traps must be visited frequently to release innocent bystanders and refresh the bait and should be labelled with the purpose of the trap, directions not to touch and a contact number of the person responsible for looking after them.

Pros and cons

May be effective with sufficient access, time and resources to cover a large area.

Monitoring

Threat Level: All

Aim. To identify the presence of Asian Hornets (including emerging queens plus migrating, hitch-hiking or wind-blown individuals of all castes).

Method. Stations should be placed in locations away from the apiary where surveillance is easy, visually or by camera, and bait can be refreshed. For example, a monitoring station outside the kitchen or office window is an ideal place.
Wick or dish bait stations should be used. If traps are deployed instead, they must be of a type that minimises by-catch.
Monitoring can be carried out from late February until late Autumn.

If Asian Hornets are seen, photograph and record them with the Asian Hornet Watch App to alert the NBU and contact your local Area/County Asian Hornet Coordinator. Enhanced surveillance in the area or track and trace operations may then be put in place.

Pros and cons

Potentially a cost-effective early warning system without by-catch if an open bait station is used.

If a closed trap with escape holes is deployed, bycatch will be reduced but not excluded.

Bait stations

Threat level All.

Aim To attract Asian Hornets for monitoring, and for track and trace operations.

Method The bait station is open and available for the Asian Hornets to visit freely. During track and trace operations they are then marked, either while feeding or briefly captured, before returning to the nest.

Usually very simple, bait stations are made from everyday household objects such as jam jars or take-away boxes and bait-soaked sponges or J-cloths. This allows visiting hornets to feed without drowning or having to stop and clean their feet which might affect the direction of departure from the bait station to the nest.

Pros and Cons

Cost effective, multi-purpose and avoids by-catch.

Asian Hornets on bait station
Courtesy Marc Struye (Belguim)

'Decoy' trapping during predation in the apiary

Threat level. Red.

Aim. To divert hornets and so reduce the level and stress of predation.

Method. Selective traps are placed near the hives to divert hawking hornets from the foraging bees. Decoy traps should only be set when *predation has actually started*, otherwise they increase risk by attracting hornets to the apiary.

The traps should be placed at least 1.5 metres from the nearest hive to prevent the Asian Hornet pheromones from stressing the bees.

A typical layout would be:

Apiary layout with 'decoy trap"

Note: hives grouped together for security.

Pros and Cons.

Decoy trapping can reduce stress on the bees but is unlikely to have any impact on the number of nests the following year.

Autumn Trapping

Threat level: amber and red

Aim. To catch newly mated queens before they hibernate and so reduce the number of foundress queens that establish new colonies the following Spring.

Method. An Autumn trapping campaign requires the same resources and effort as in Spring.

Pros and Cons. Autumn trapping is the most favoured by those wishing to minimise by-catch because there are fewer other species on the wing.

However, there seems to be no evidence that Autumn trapping makes any difference due to the high natural attrition rate of hibernating queens and is the least popular. If traps using pheromones as bait become available, trapping during the mating period could become more effective.

Trap Design

Trap design is still developing and there is much debate about what works best.

Views may change as the UK situation progresses, but naturally we all want to be proactive and source the best equipment and techniques available to protect our bees.

Current thinking is that traps should have a **7-8 mm diameter entry** which is large enough to admit and retain the Asian hornet, but small enough to exclude bigger insects (e.g. European hornets and butterflies). The **exits should be of 6.5 mm,** at the highest point of the trap, to let as many smaller, non-target species escape as possible.

Popular traps are the 'grill and cone type 'such as the *Jabeprode* or *Gard'Apis*, whereas bottle-shaped designs may be less effective in reducing by-catch. The *Gard'Apis* trap has two different sized entrance nozzles: the 8 mm red nozzle is designed to catch queens in the Spring and Autumn; and the 7 mm orange nozzle for workers in Summer.

Gard'Apis

Jadeprode

Other, less expensive, traps such as the VetaPharma are useful when adapted to ensure there are 6.5 mm escape holes at their highest point. A new model designed to allow bycatch to escape is expected to be available in the UK shortly.

VitaPharma trap

More versions are available commercially, and some are being produced by individual or groups of beekeepers who have access to 3-D printing.

Traps should be clearly marked to show their purpose, instructions not to touch, and the contact details of the person responsible for the trap.

The Great Bait Debate

The ideal bait is one that attracts Asian Hornets but is repulsive to all other species. Despite extensive research in Europe and Asia, no such bait has yet been identified, although trials with pheromones are looking promising.

In France a mixture of equal parts of pressed fruit juice, beer and dry white wine is widely used. For the cost-conscious, fermented honey or wax cappings seem also to be enjoyed by the hornets. On Jersey the commercial wasp attractants produced by *Trappit* have been effective in all seasons. Liquid bait should be soaked in a sponge or pot with a wick to prevent other insects from drowning.

But by far the best attractant is the smell of other hornets. Once there are hornets in the traps, even after they are dead, they are a powerful lure to others.

Disposal of Trap Contents

Traps should be emptied after dusk when Asian Hornets have stopped flying. Put the trap in a freezer for 30 minutes to stun the contents and release any bycatch. Using a spoon, transfer all but 3 Asian Hornets to a plastic bag and return it to the freezer for 48 hours to kill them. The remaining 3 Hornets can be returned as bait to the unrinsed trap ready for redeployment.

Recording and mapping

In addition to the normal hive records, it is essential to keep notes about traps and bait stations: locations; date/times of checks; trap type; bait type; catch; actions taken.

Mobile apps such as Epicollect, Google My Maps and the Hampshire AH Tracker are ideal methods for recording and sharing information with the NBU, local AHATs and other beekeepers about the location of monitoring stations, sightings, nests etc.

At the time of writing the BBKA/BDA* eR2 record and mapping system is being extended to include the location of AHATs and the position of traps. The system described can be used to ensure that the traps are not disturbed by other beekeepers or members of the NBU.

*Bee Diseases Insurance Ltd.

Finding and Destroying Nests

"To reduce the amount of predation you have to find and destroy the nests."

Alistair Christie, Asian Hornet Coordinator States of Jersey

This is not something beekeepers are likely to be involved in unless they are members of an AHAT, so this chapter gives a brief outline of the procedures.

Although nest destruction alone is unlikely to solve the Asian Hornet problem once a population is established, it can help to slow down the speed of the invasion, reduce predation on apiaries and relieve the pressure on biodiversity in the vicinity.

You can never be sure to have found all the nests, and other new insects may be arriving all the time either migrating, blown on the wind, or as hitchhikers.

Asian hornet nest removal is a dangerous operation and is a job for properly trained and equipped professionals.   On no account should you approach a nest or try to remove it yourself.

Whilst the hornets are not usually aggressive when out hunting, they become very dangerous to anyone approaching too closely to or disturbing the nest. They can attack in large numbers and each inflict multiple, deep and potentially fatal stings.

Track and trace

The most effective way of locating nests is to **catch, mark and release** hornets using bait stations. A bait station is located in an area where Asian Hornets are believed to be flying. The hornet is attracted to the bait and, while feeding, may be marked to allow for later identification. When the hornet flies off, its direction is plotted using a compass. Its departure time is also recorded so that, when the identifiable hornet returns to the bait station, the total intervening time can be calculated. Recognizing that Asian Hornets fly at a rate of 100 metres per minute and are likely to spend 30 seconds in the nest before returning, the approximate distance to the nest can be calculated. Combining the distance calculation plus the direction gives a line to the nest. This process is repeated at two other bait stations, then moved down the flight line until close to the nest to triangulate and pinpoint its position.

The hornets are marked with a queen marker pen or may have a tinsel ribbon glued to the thorax. If tinsel is used the queens are also marked with coloured pens on the abdomen. Experiments using miniature radio transmitters attached to the hornets may bring more speed and accuracy in the future.

Asian Hornet on a wick bait station
Photo by Alan Baxter

Asian hornets on an open bait station
The stone enables feeding without getting their feet wet.

Radio tagging in Belgium. Smaller tags are currently being developed.
Courtesy of Marc Struye (Belgium) ©

At the time of writing, invasive species can only be caught and released by licensed operators, in the case of the Asian Hornet this is limited to members of the National Bee Unit. It is anticipated that suitably trained AHAT volunteers may be licensed to carry out catch, mark and release operations in the future.

Nest Removal in Southampton

Following a report from a member of the public, track and trace operations by the NBU revealed a nest in the Old Cemetery in Southampton. Hidden in a tall tree surrounded by undergrowth it was necessary to deploy tree climbers to access the nest for destruction and removal. The nest and contents were sent to the Fera Laboratory for analysis. The following images give an idea of the difficulty of the task.

The tree.

Closer view of the tree.

Preparing and loading the lance.

Climbing the tree.

Killing the nest with the lance.

Cutting off the branch and bagging the nest.

Dan Etheridge with nest.
All photos by Alan Baxter

Conclusion

After a long 'phony war' it seems the much- feared arrival of the Asian Hornet on our shores has finally come to pass. Suddenly beekeepers are faced with new challenges that may at first seem overwhelming.

However, all is not lost. By accepting that we have to live with the Asian Hornet, deciding we will not be beaten by it, and by understanding more about it, we can learn how to manage the problem.

This may mean that we have to sharpen up our beekeeping skills and adapt our beekeeping methods to mirror the life cycle of the hornet.

With a phased, targeted response to the threat and by adopting the 3 pillars of Fit2Fight:

▶ Healthy bees

▶ Strong colonies

▶ Well-fed stocks

We can manage the situation, reduce the level of predation and relieve the stress of predation on our apiaries.

With a positive attitude, and using the simple weapons at our disposal, we can continue to enjoy our beekeeping, achieve good outcomes for our bees, and look forward to sharing the wonderful produce of the hive.

FIT2FIGHT ANNEXES

Annex A Instructions for Shook Swarm

Equipment

- ▶ Clean floor and entrance blocker

- ▶ Clean brood box

- ▶ 11 frames of foundation

- ▶ Dummy board

- ▶ Clean crown board

- ▶ Roof

- ▶ Rapid feeder

- ▶ 4 litres of thick sugar syrup

- ▶ Queen cage

- ▶ Bin bags for the old frames of comb

Method

- ▶ Move the hive to one side.

- ▶ Put a clean floor with entrance blocker on the original stand. The flying bees will return to this new brood chamber.

- ▶ Ensure the entrance blocker is in the closed position.

- ▶ Place a queen excluder on the floor to act as a queen includer, preventing the bees from absconding with the queen.

- Add a fresh brood box of foundation.

- Remove four frames from the middle of the new brood box to create a space.

- Dismantle the original hive.

- Find and cage the queen for safe keeping.

- Remove each brood frame in turn and shake the frame into the space in the centre of the new brood box.

- Gently fill up the brood box with the spare frames.

- Release the queen into the new brood box.

- Add the rapid feeder with thick sugar syrup.

- Close the hive.

- Scrape and disinfect the old brood box and other equipment.

- Destroy the old comb by burning.

- After about a week check that the queen is laying and if so, remove the queen includer.

- Keep feeding until all the comb has been drawn out. You may need to turn round the end frames as the bees can't draw out frames next to the hive wall.

- Any supers can be put on top of the feeder until it's no longer needed.

NB Shook swarm should only be performed in spring on strong colonies with a young queen.

Annex B Instructions for Bailey comb change

This is carried out as follows:

Day 1

- ▶ Place a brood box containing frames of foundation directly on top of the original retaining

- ▶ Add a feeder of thin sugar syrup unless there is a strong flow in progress to stimulate the bees to draw out the foundation.

Day 8

- ▶ Close the lower entrance.

- ▶ Add a Bailey board, or a queen excluder with an eke and an entrance, between the brood boxes to confine the queen in the upper box if she has started to lay in it.

- ▶ Top up the feeder if required.

About Day 21- 24

- ▶ Once the foundation has been fully drawn remove the feeder, add a queen excluder and replace any supers.

- ▶ Place the new brood box on a clean floor.

- ▶ Remove the old comb for rendering down or destruction.

- ▶ Scrape and scorch the old frames and brood box for reuse, or disinfect poly equipment with a strong solution of soda crystals and bleach.

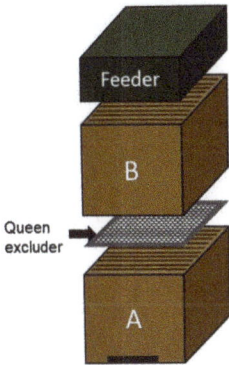

DAY 1:
Place box of foundation and
feeder above colony

A: Old comb, queen & brood
B: Brood box of fresh
foundation

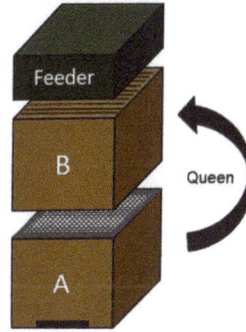

Day 7:
Queen is taken from box A
and placed into box B

A: Old comb & brood
B: Box of partly drawn
foundation & queen

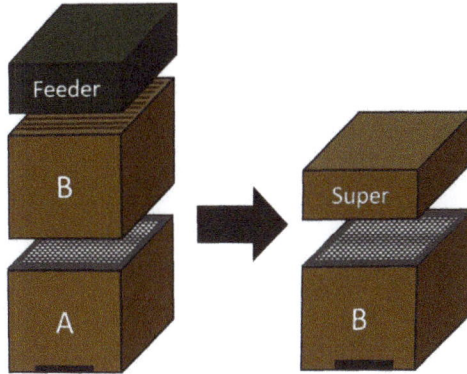

Day 21-24:
When all brood has emerged from box A, it
can be removed. The colony is
reassembled

A: Old comb; empty
B: New comb, queen & brood

Bailey comb change for a strong colony

125

Annex C Testing for Nosema

Equipment

- ▶ Mortar & pestle
- ▶ Scissors and forceps
- ▶ Distilled water
- ▶ Pipette
- ▶ Slides and cover slips
- ▶ Compound microscope x 400

Method

- ▶ Remove the abdomens of 10 bees by snipping them at the petiole (waist)and place in the mortar.
- ▶ Add 5 ml of distilled water.
- ▶ Crush the abdomens thoroughly with the pestle.
- ▶ Using the pipette, place about 10ml of the soup on a slide.
- ▶ Put on the cover slip.
- ▶ Examine under microscope at 400 x magnification.
- ▶ Nosema spores are seen as rice-grain shapes:

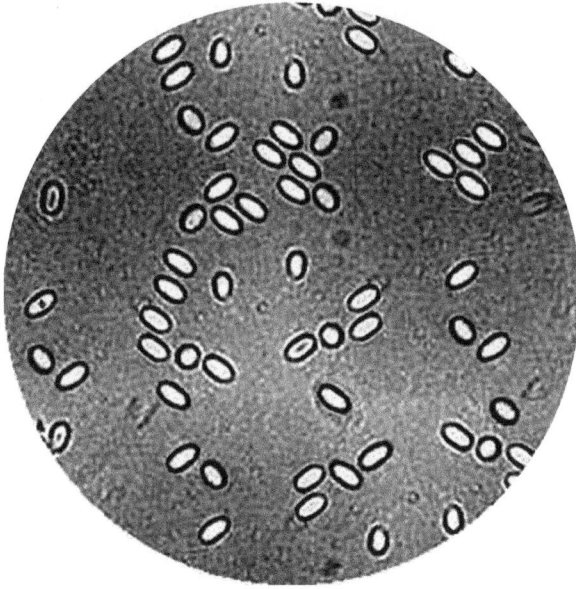

Nosema spores at x 400 magnification

Annex D Bailey comb change for a weak colony

Equipment:

- ▸ Clean brood chamber containing frames of sterilised drawn comb
- ▸ 4 Dummy boards
- ▸ Queen excluder and eke with entrance block or a Bailey Board
- ▸ Crown board
- ▸ Floor
- ▸ Contact feeder and thick sugar syrup.

Method:

Day 1

- ▸ Take the frame with the queen on and place it in the new brood box.
- ▸ Remove all frames without brood from original brood box and any supers.
- ▸ Close up the remaining frames in the centre of the brood box with dummy boards.
- ▸ Put frames of new drawn comb in the new box on either side of the frame with the queen.
- ▸ Match the number of drawn combs with the number in the bottom box.
- ▸ Close up the space in centre of the new box with dummy boards.
- ▸ Put a queen excluder and an eke with an entrance block (or a Bailey Board if you have one) on the old brood box.
- ▸ Place the new brood box with clean drawn frames above the lower brood box.
- ▸ Put on the clean crown board.
- ▸ Feed with thick sugar syrup.
- ▸ Add the roof.
- ▸ Dispose of the old comb by burning or rendering down.
- ▸ Scrape and disinfect by scorching the old frames.

Day 7

▸ The queen in the upper box should have moved onto the new frames and started to lay.

▸ Remove the old frame with the remaining brood and return it to the bottom box.

▸ Add more drawn sterilised frames to the new box and close up with dummy boards.

▸ Check lower box for queen cells and remove if found.

▸ Remove any frames that have no brood.

▸ Dispose of the old comb by burning or rendering down.

▸ Continue to feed.

Day 7-28

▸ Add a few more drawn sterilised frames in the top brood box.

▸ Remove frames with no brood from the bottom box.

▸ Dispose of the old comb by burning or rendering down.

▸ Replenish the feed.

Day 28

▸ Place a clean floor on the stand and add an entrance blocker.

▸ Put the upper box on the new floor and adjust the entrance.

▸ Add a queen excluder then the super(s).

▸ Close the hive with the clean crown board and the roof.

▸ Remove, scrape and disinfect the old queen excluder and upper entrance or Bailey Board, lower brood box and old floor.

▸ Dispose of old comb by burning or rendering down.

BAILEY COMB CHANGE FOR A WEAK COLONY

A method of transferring a weak colony of bees onto
new drawn comb without loosing brood and causing as little stress as possible

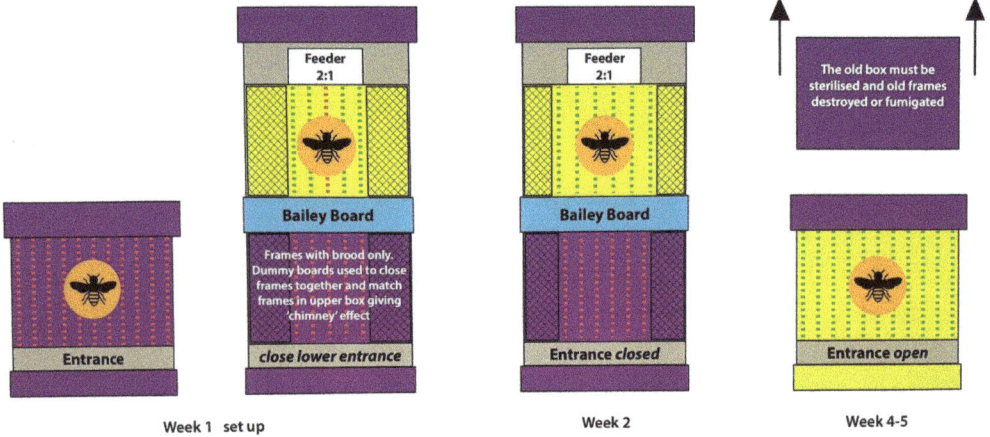

Feeder 2:1

Bailey Board

Frames with brood only. Dummy boards used to close frames together and match frames in upper box giving 'chimney' effect

Entrance

close lower entrance

Week 1 set up

Feeder 2:1

Bailey Board

Entrance *closed*

Week 2

The old box must be sterilised and old frames destroyed or fumigated

Entrance open

Week 4-5

Bailey comb change for a weak colony

Annex E Nucleus Method of Swarm Control

Equipment needed

▸ Nucleus box

▸ 4 frames of drawn comb

▸ Feeder (some models may have integral feeders)

▸ Feed of thin sugar syrup

Method

Day 1

▸ Place the nuc beside the parent hive.

▸ Find the frame with the queen and carefully place it in the nuc. **This frame must have no queen cells.**

▸ Add another frame of brood in all stages.

▸ Add a frame of liquid stores.

▸ Add a frame of pollen.

▸ Shake in 3 frames of young nurse bees (tap the frames gently to dislodge any older bees).

▸ Move the nuc at least 2 metres away with the entrance facing in a different direction.

▸ Reduce the nuc entrance to the queen excluder setting to prevent the queen from absconding.

▸ If there is no nectar flow, add the feeder with thin sugar syrup to stimulate the queen to resume laying.

▸ Mask the entrance with grass to force the bees to reorientate to their new location.

▸ Inspect the parent hive frame by frame, shaking or brushing off all the bees. Don't shake the frames with queen cells to avoid damaging the larvae inside.

▶ Select two uncapped queen cells with healthy-looking larvae and plenty of royal jelly, marking the frames with a drawing pin.

▶ Break down all other queen cells.

▶ Replace frames that you removed to make up the nuc with frames of foundation or drawn comb.

▶ If there is no flow, add a rapid feeder with thin sugar syrup to help the house bees feed the queen larvae.

Day 2

▶ Check that the nuc still has enough young bees (some may have left and returned to the parent).

▶ Shake in another frame of young bees if necessary.

Day 8

▶ Check the nuc and the parent for new queen cells and break down any that are found.

▶ In the parent hive remove one of the two queen cells that you have saved.

▶ Top up feed if needed.

▶ Leave undisturbed for 3 weeks to allow the new queen to emerge and go on her mating flights.

Equipment needed:
- Clean floor
- Brood box
- Cover board
- Roof
- 11 Frames of drawn comb or foundation
- Dummy board
- Extra queen excluder
- Rapid feeder
- Thin sugar syrup
- Drawing pin

Method
- Move the hive to a new stand at least 1.5 metres away. This will be **'the parent'.**
- On the original stand, place a new floor and a queen excluder.
- Add the new brood box of foundation or drawn comb and the dummy board.
- Remove the 3 middle frames and put to one side.
- This new box will be **'the swarm'.**
- From the **parent**, take the frame with the queen and place in the centre of the **swarm. This frame must have no queen cells.**
- Add a frame of mainly capped brood.
- Add a frame of stores.
- Add a queen excluder, the supers, crown board and roof.
- Replace the missing frames in the parent.

▸ Check the **parent** for queen cells and select two well-provisioned uncapped queen cells. Mark the frame with a drawing pin.

▸ Shake off the remaining frames and break down all other queen cells.

▸ Check for stores and add a feed of sugar syrup if needed (there will be no foragers bringing food into this colony yet).

▸ Put on the crown board and the roof.

Day 8

▸ Check both hives for queen cells and break down any new ones that you find.

▸ In the **parent** remove one of your selected sealed queen cells that will now be sealed.

▸ In the **swarm** remove the queen excluder from under the brood box if the queen has started to lay.

▸ Check both colonies for stores and top up feeders if required.

▸ Do not disturb the **parent** during the day for 3 weeks to allow the new queen to mate.

Day 21

Check the **parent** to see if the new queen is laying

▸ You may now choose to keep both colonies or remove one queen and merge them into one large colony.

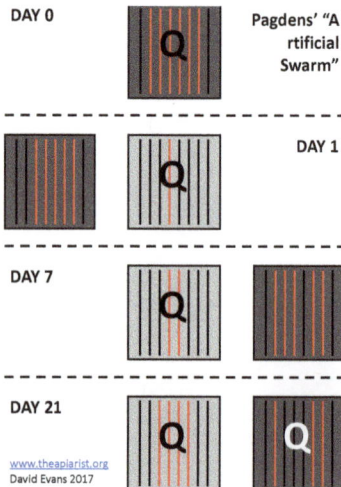

Pagden Method of artificial swarm
Copyright David Evans 2017

Annex G Demaree Swarm Control

If you don't want to increase the number of colonies in your apiary or you are short of space, this very simple method is the one for you.

Equipment needed:
- Clean brood box
- 11 frames of drawn comb or foundation
- 2 x supers
- 2 x Queen excluders
- 1 x dummy board

Method:
Day 1
- Move the brood box to one side.
- Place the new brood box on the original floor.
- Add 9 frames of drawn comb or foundation leaving a gap in the middle.
- Go through the original box and find the frame containing the queen.
- This frame must have **no queen cells.**
- Put this frame with the queen in the middle of the new box.
- Add the extra frame and dummy board.
- Add a queen excluder.
- Put on 2 supers above the QE.
- Add the second QE.
- Put the original brood box with the brood on top.
- Inspect the original brood box, shake off all frames & **destroy any queen cells.**
- Add a frame to replace the one with the queen.

- ▸ Put on crown board and roof.

- ▸ Leave for a week.

Day 8

- ▸ Inspect the top box and **remove any queen cells.**

- ▸ Close the hive and wait for all the brood in the top box to emerge.

- ▸ Leave for about 3 weeks.

Day 25

- ▸ All the brood in the upper brood box has emerged.

- ▸ Remove upper brood box and reassemble the hive.

Demaree swarm control

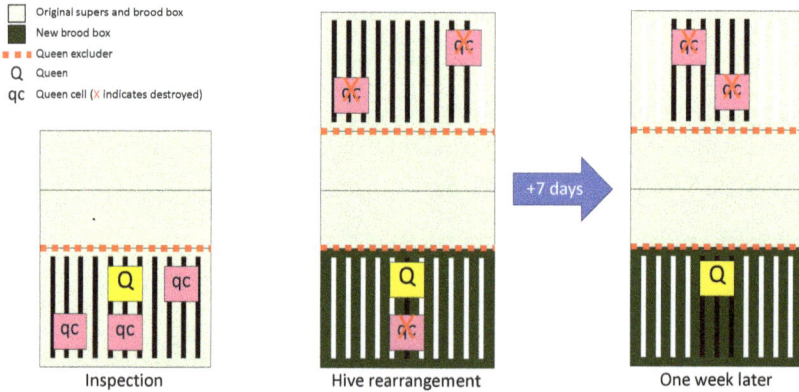

- □ Original supers and brood box
- ■ New brood box
- ▪ ▪ ▪ Queen excluder
- Q Queen
- qc Queen cell (X indicates destroyed)

Inspection   Hive rearrangement   One week later

+7 days

@David Evans 2019
www.theapiarist.org

Pagden Method of artificial swarm
Copyright David Evans 2017

Annex H Winter Preparation

What follows was published as a blog in August 2024 and has already been briefly described in previous chapters, but in view of the degree of importance of winter preparation in the fight against the Asian Hornet it is repeated in full here.

Introduction

As we've seen, the honey bee colony builds up to its maximum foraging force in July for the summer flow when they must accumulate enough food stores to last for the next 7 or 8 months. The population of 'summer' bees rapidly declines from August onwards until there only remains a small number of 'winter' bees. Unlike summer bees, who only live for 5 or 6 weeks, the winter bees can live for up to 6 months. Their job is to keep the queen alive and ready to start producing brood again in a few months' time, as well as to feed the spring larvae. Colonies can be lost in the spring when not enough winter bees have survived to keep the colony alive. The more winter bees, and the longer they live, the greater the chance of the colony surviving into the following year.

Bees emerging in August and September will die in October and November. Bees born in early October are those that can live until March IF there are enough of them and they are healthy, have a low varroa load, enough food and well-provisioned fat bodies.

Feeding

Stimulating the queen to lay

Since the bees that are born after the beginning of October will live until next spring, it is vitally important that the queen continues to lay after the summer peak to produce plenty of heathy, well-fed winter bees. This can be achieved by stimulative feeding of **thin 1:1** sugar syrup, and pollen patties if there is little pollen coming in, in mid to late September to simulate a nectar flow and encourage the queen to keep laying. Note that **thick 2:1** syrup or Inverted syrup will only be laid down as stores.

Winter stores

A strong colony is estimated to need about 21 kg of honey to avoid starvation. Happily for us, the bees usually collect more than they need and the surplus provides us with the delicious honey that we enjoy so much. If we take more, or if the bees haven't collected enough, we must replace it with some form of artificial feeding, usually in the form of sugar syrup, pollen supplement and fondant.

Hefting the hives regularly in winter can tell us if they have enough. If the hive feels light, feeds of **thick 2:1 syrup** can be given if the daytime temperature is above about 10 deg C. Once the temperature falls below this level the bees will be reluctant to move upwards to feeder and will be unable to digest the cold syrup, so **fondant** is required.

Some beekeepers leave the bees with a full super. It's important to remove the queen excluder so the queen isn't isolated when the cluster moves into the super. Whether the super is placed under or over the brood box doesn't seem to make any difference, although bees will often prefer to move upwards, so it could be argued that above is better. In some cases, the cluster will not move up and over the top of the frames to reach the stores in nearby frames and the colony dies in what is known as isolation starvation. Communication holes made in the comb can help the cluster to move more easily from one frame to the next.

Health

A thorough health inspection after the supers have been taken off gives a warning of any adult or brood diseases which can then be treated, hopefully in time. See Chapter x for more information on pests and diseases.

Varroa

Varroa infestation is one of the main causes of winter losses and must be treated as soon as the honey harvest has been taken off.

Although the bee population is decreasing in autumn the varroa load continues to rise exponentially. The untreated colony can soon be overwhelmed, with fatal results. Apart from weakening the adult bees and shortening their lives, Varroa can cause them to suffer from a variety of serious virus infections including Deformed Wing Virus (DWV), Chronic Bee Paralysis Virus (CBPV) and more. There are numerous approved products on the market, some containing natural ingredients such as Formic Acid, Thymol and essential oils. Others are composed of manufactured chemicals including Amitraz, Flumethrin and Tau Fluvalinate. Artificial or 'hard' chemicals are effective but carry the risk of the varroa developing resistance.

Whilst Formic Acid and Thymol treatments are said by some to be tough on the queens or cause excessive deaths of workers, all the UK Veterinary Medicines Directorate (VMD) authorized brands are safe to use if the manufacturer's instructions are followed to the letter, especially regarding dosage, temperature and ventilation. All are designed to kill the mites that are feeding and reproducing inside the capped brood.

Treatment must be applied after removal of the supers containing honey for human consumption. Any supers left in place during treatment should be marked to indicate that they may contain traces of the miticide used and should not be used again until the frames have been cleaned and refitted with new foundation.

A second treatment of Oxalic Acid to kill the 'phoretic' mites, i.e. those living and feeding on the adult bees, is given in winter when there is little or no brood. Traditionally this was carried out in late December between Christmas and New Year, but recent research has shown that in Southern England the optimum period is now early December. Treatment is by trickling the dissolved product from a syringe, using a device called Gas Vap which is a modified blow torch, or by sublimation with special equipment. Treatment with Oxalic Acid requires the operator and any bystanders to wear personal protective equipment.

Remember that all chemical treatments must be recorded on Form VMD 5 which is downloadable from the VMD Website.

Queens

Apart from Varroa and its attendant diseases, queen failure is probably the second most common cause of winter losses. Queens are at their most productive in the first 2 years of their life. After that their egg laying capacity declines and they are likely to be superseded. Unfortunately, the chances of a replacement queen being mated are diminishing rapidly and both the old and the new queen will fail to survive.

Autumn is a good time to introduce new, mated queens which are ready to produce the vital winter bees needed to see the colony through winter and into spring. If the old queen has good genetic traits, and you can't bear to part with her, she can be retired into a nuc and used for emergencies such as making up winter losses, or for raising new queens next year.

Young queens in their first 2 years are less inclined to swarm the following year. Those that do make swarm preparations will pass on the trait to their offspring and their drones will spread the swarmy behaviour to other colonies in the area. The colony should be requeened as soon as practicable.

Similarly, if the colony has been excessively defensive or flighty during the season, requeening now will improve the temperament of the colony and make your beekeeping a pleasure again, as well as not spreading the undesirable genes to other beekeepers' stock by its drones when mating with their queens.

Like humans, not all queens are born equal and poorly mated or badly developed queens will fail early in their lives or be superseded.

Wax Moth

Wax Moth can inflict serious damage to stored brood boxes and supers, rendering the comb unusable and often leaving the frames damaged.

Treatment by freezing at -18 deg C for 48 hours will kill any eggs or larvae of Wax Moth and *Braula coeca* (Bee Louse).

Alternatively, fumigation for a week with 80% acetic acid will do the job. Be careful to follow the manufacturer's instructions regarding safety and remember to protect any metal parts with Vaseline and not to stand the stack of boxes direct onto concrete.

FIT2FIGHT CALENDAR

Approximate timing of activity depending on location and weather

MONTH	AH ACTIVITY	ALERT STATE	ACTION IN APIARY
January	Queens in hibernation	Green	Check stores and add fondant if necessary
February	Depending on weather queens emerge from hibernation & build embryo nest	Green	Feed pollen & syrup to stimulate queen to lay Move hives closer together, mark entrances to reduce drifting. Put out monitoring stations Report sightings on AH Watch App
March	Queens continue to emerge. Building primary nests First workers emerge	Green	First spring inspection Remove mouse guards, change floors Health inspection Check varroa loads and treat if >1,000 per hive Maintain monitoring stations Feed pollen & syrup to encourage queen laying and worker wax gland development for comb building. Unite weak colonies. Inspect weekly for growth and swarm prevention. Consider shook swarm depending on weather.
April	Queens continue to emerge	Amber	Health inspection. Maintain monitoring stations Start queen rearing preparations Carry out splits to make increase if required Equalize colonies for simplicity of management Inspect weekly for growth, take samples for Nosema testing if growth is slow. Swarm prevention Ensure queens have enough room to lay Move full frames to outside and empty frames near centre Add supers
May	Queens continue to emerge.	Amber	Queen rearing begins Inspect weekly for growth and swarm prevention. Add extra brood box to give more space for queen to lay and bonus of getting drawn comb Add more supers Maintain monitoring stations Let grass grow round hives or fit boards round sides
June-July	Rapid growth of colony. Building secondary nest	Amber	Queen rearing ends. Maintain monitoring stations Inspect weekly for growth and swarm prevention. Move mating nucs to full brood box or 14x12 nucs. Unite weak colonies. Feed if no flow Drone brood uncapping for varroa control Remove honey crop, leave min 1 full super and a half to give bees room Treat for varroa.
Mid July	Predation period starts Colony building to maximum strength	Red	Final manipulations completed. Reduce entrances to 1 beespace only if predation taking place Feed syrup, pollen, water early morning or evening Provide shade if needed Put woven tangles of branches in front of hives or fit muzzles and deploy harps if you have them 'Decoy' Traps only during predation Open entrances at night
August	Predation period	Red	Feed syrup, pollen, water early morning or evening Keep entrances closed to 1 beespace only when predation in apiary Provide shade if needed Decoy Traps only when predation in apiary Open entrances overnight
September	Predation period	Red	Feed syrup, pollen, water early morning or evening Keep entrances closed entrances to 1 beespace only when predation in apiary Provide shade if needed Decoy Traps only when predation in apiary Open entrances at night
October	Predation period Sexuals produced Mating takes place	Red	Feed syrup, pollen, water early morning or evening Open hive entrances if all clear. Open entrances at night Fit mouseguards Consider green woodpecker prevention Fit insulation if not already done
November	Colony and old queen die Nest abandoned. New queens to hibernation	Amber	Open hive entrances if all clear. Heft and feed foundation if necessary
Deecember	Queens in hibernation	Green	Treat with Oxalic acid. Heft and feed foundation if necessary

Note: All timings applicable to the South of England. Other regions may be later.

141

Abbreviations

AH	Asian Hornet
AHAT	Asian Hornet Action Team
AH Coord	Asian Hornet Coordinator
APHA	Animal and Plant Health Agency
BBKA	British Beekeepers Association
BDI	Bee Diseases Insurance
CBPV	Chronic Bee Paralysis Virus
DWV	Deformed Wing Virus
MAQS	Mite Away Quick Strip
NBU	National Bee Unit
OA	Oxalic Acid
OSR	Oil Seed Rape
VMD	Veterinary Medicines Directorate
YLAH	Yellow-legged Asian Horne

Source Material/References

Basterfield D. Davis I. (2019) British Beekeepers Association

Baxter A. https://www.alanbaxtersblogs.co.uk

BBKA Healthy Hive Guide

BBKA Special Edition: *Feeding Honey Bees*

Bunker S. '*The Yellow-Legged Asian Hornet - Biology, Spread and Control*'. (Published 2024).

Durham A. https://shorturl.at/dnqG9

Gregory P *Healthy Bees are Happy Bees* 2nd Edition 2028 Bee Craft Ltd

Evans D the Apiarist

Inventaire National du Patrimoine Naturel INPN https://shorturl.at/htXY8

Mitchel D. University of Leeds https://shorturl.at/dmwAS

National Bee Unit Queen trapping guidance https://shorturl.at/8zN2R
National Bee Unit Managing Varroa https://shorturl.at/ysjpR

Pederson, Kennedy et al. Broad ecological threats of an invasive hornet revealed through deep sequencing approach. Science of the Total Environment 4 March 2025. https://www.sciencedirect.com/science/article/pii/S0048969725006138

Shaw W. *Feeding Bees* Northern Bee Books 2021

Stainton, K. (2022) *Varroa Management: A Practical Guide on How to Manage Varroa Mites in Honey Bee Colonies*, Northern Bee Books.

Acknowledgements

With grateful thanks to Ali Everest, Sarah Bunker, and Helen Tworkowski for their combined wisdom, and for pointing out my numerous errors and shortcomings.

And especially to my most exacting critic, my wife Penny Melville-Brown OBE DUniv.

Any mistakes or omissions in this book are entirely my own.

Writing a book about bees can be a minefield. There's an old beekeeping saying:

"If you ask 5 beekeepers a question you get 6 different answers."

How many answers do you get if you ask 3 beekeepers and a lawyer?

9 781914 934995